Advances in Planetary Science – Vol. 4

PLANETARY HABITABILITY IN BINARY SYSTEMS

Advances in Planetary Science

Print ISSN: 2529-8054
Online ISSN: 2529-8062

Series Editor: Wing-Huen Ip *(National Central University, Taiwan)*

The series on Advances in Planetary Science aims to provide readers with overviews on many exciting developments in planetary research and related studies of exoplanets and their habitability. Besides a running account of the most up-to-date research results, coverage will also be given to descriptions of milestones in space exploration in the recent past by leading experts in the field.

Published

Vol. 4   *Planetary Habitability in Binary Systems*
by Elke Pilat-Lohinger, Siegfried Eggl and Ákos Bazsó

Vol. 3   *Planetary Habitability and Stellar Activity*
by Arnold Hanslmeier

Vol. 2   *Origin and Evolution of Comets:*
Ten Years after the Nice Model and One Year after Rosetta
by Hans Rickman

Vol. 1   *Nuclear Planetary Science: Planetary Science Based on Gamma-Ray,*
Neutron and X-Ray Spectroscopy
by Nobuyuki Hasebe, Kyeong Ja Kim, Eido Shibamura and
Kunitomo Sakurai

Advances in Planetary Science – Vol. 4

PLANETARY
HABITABILITY
IN BINARY SYSTEMS

Elke Pilat-Lohinger
University of Vienna, Austria

Siegfried Eggl
Jet Propulsion Laboratory (JPL), USA

Ákos Bazsó
University of Vienna, Austria

World Scientific

NEW JERSEY · LONDON · SINGAPORE · BEIJING · SHANGHAI · HONG KONG · TAIPEI · CHENNAI · TOKYO

Published by

World Scientific Publishing Co. Pte. Ltd.
5 Toh Tuck Link, Singapore 596224
USA office: 27 Warren Street, Suite 401-402, Hackensack, NJ 07601
UK office: 57 Shelton Street, Covent Garden, London WC2H 9HE

Library of Congress Cataloging-in-Publication Data
Names: Pilat-Lohinger, Elke, author. | Eggl, Siegfried, author. | Bazsó, Ákos, author.
Title: Planetary habitability in binary systems / by Elke Pilat-Lohinger
    (University of Vienna, Austria), Siegfried Eggl (Jet Propulsion Laboratory (JPL), USA),
    and Ákos Bazsó (University of Vienna, Austria).
Description: New Jersey : World Scientific, [2018] | Series: Advances in
    planetary science ; volume 4 | Includes bibliographical references and index.
Identifiers: LCCN 2018040142 | ISBN 9789813275126 (hc : alk. paper)
Subjects: LCSH: Habitable planets. | Double stars.
Classification: LCC QB820 .P52 2018 | DDC 523.2/4--dc23
LC record available at https://lccn.loc.gov/2018040142

British Library Cataloguing-in-Publication Data
A catalogue record for this book is available from the British Library.

For any available supplementary material, please visit
https://www.worldscientific.com/worldscibooks/10.1142/11125#t=suppl

Desk Editors: Anthony Alexander/Amanda Yun

Typeset by Stallion Press
Email: enquiries@stallionpress.com

This book is dedicated to our dear families. Their continuous support and kind encouragements were vital for us in the past, and hopefully will last for many years to come.

E. P.-L.: Carolin, Fabian and Johannes
S. E.: Christine, Hartmut and Siegfried
A. B.: Julianna, Ferenc and Botond

Preface

The search for planets outside the Solar System revealed an unexpected diversity of planetary systems. Strange new worlds, such as pulsar planets, hot Jupiters, planets in binary star system and planets in highly eccentric orbits have been found.

Detections of such substellar companions, particularly in multiple star systems, have led to a growing interest in studies of binary star–planet configurations. Observational surveys yield a relatively high percentage of multiplicity among Sun-like stars in the solar neighborhood. This is the region where space missions like CHEOPS, TESS or PLATO are most effective in their search for habitable planets outside the Solar System.

In order to assess the habitability of planets in binary star systems, not only are astrophysical considerations regarding stellar and atmospheric conditions needed, but orbital dynamics and the architecture of the system also play an important role. Due to the strong gravitational perturbations caused by the presence of the second star, the study of planetary orbits in double star systems requires special attention. This monograph focuses on the gravitational interactions of celestial bodies and their influence on the habitability of an Earth-like planet in this system. Rather than presenting a broad overview on the subject, which would require a volume in its own right, this book is intended to discuss the most recent progress in the field. This book is, thus, addressing researchers and graduate students interested in the synthesis of Astrobiology and dynamics of exoplanets. To start with, we provide in the first three chapters a general overview on the basic requirements for planetary motion in binaries. The following chapters include the theoretical foundations to construct numerical codes that allow the application of the methods described in this book.

We summarize in Chapter 1 various binary star configurations known from the observations and present the most important catalogues of binary stars. With this data we show the related basic population statistics, which is relevant for dynamical studies concerning habitability.

In Chapter 2, we first discuss different types of planetary motion in binary star systems. Then we elaborate on the requirements for stable planetary motion, considering circular and eccentric orbits for both the binary and the planet. This chapter introduces the basic notions for dynamical studies.

Chapter 3 describes the main gravitational perturbations (resonances) that arise in multi-body systems. Different kinds of resonances are important for the final architecture of a planetary system. To determine the locations of such perturbations, we introduce a newly developed method which is applicable in particular to planetary motion in binary stars.

To understand whether a system can host habitable worlds requires us to consider its dynamical evolution, beginning right after its formation. In this context, we review in Chapter 4 some new aspects of the early phase of planet formation, discuss the perturbations affecting forming terrestrial planets in the habitable zone and discuss the water transport by icy planetesimals as a necessary requirement for habitability.

The influence of the secondary star for habitability from a dynamical point of view is figured out in Chapter 5. Besides the dynamical constraints, specific stellar properties, like the spectral type, the activity and evolution, are also discussed.

An important part of this monograph is Chapter 6 with the self-consistent determination of the habitable zone borders in binary star systems. We present our analytical methodology that determines the borders of habitable zones in binary star systems, taking into account the combined dynamical–radiative influence of the stars. The fact that a secondary star constantly perturbs the planetary motion requires special considerations. In this context, the consequences of orbital dynamics regarding insolation are also discussed in this chapter.

In Chapter 7, we discuss real observed binary star systems that host giant planets and analyze the habitability of these systems. The goal is to determine the stable regions for planetary motion, the habitable zone borders and the gravitational perturbations that could influence an Earth-like planet in these systems. This study allows us to distinguish between binary systems with and without habitable environments. To achieve this, we apply all necessary requirements and methods to the various systems that we have introduced and developed in the previous chapters.

Finally, in Chapters 8, we conclude the new findings for habitable planets in binary stars and we address the question of whether Solar System-like configurations are needed to discover an exo-Earth and discuss such configurations in binary star systems.

We believe that this monograph is interesting as it can serve current and future astronomers searching for potentially habitable planets in the solar neighborhood and beyond.

Elke Pilat-Lohinger
Siegfried Eggl
Ákos Bazsó

Vienna/Pasadena
February 2018

About the Authors

Elke Pilat-Lohinger is currently project leader, researcher and lecturer at the University of Vienna, Austria where she received her PhD in 1994. After graduation she moved to France for a postdoctoral fellowship (FWF — Erwin-Schrödinger grant) at the Observatory of Nice in 1996 and 1999. Dr. Pilat-Lohinger has worked at the University of Vienna since 2000, with the financial support of the Austrian Science Fund (FWF). With the honourable Hertha-Firnberg-Grant, she started working on the stability of extra-solar planetary systems, followed by a series of FWF projects covering the following topics: extra-solar planetary research, astrobiology, dynamical astronomy, and solar system dynamics.

Siegfried Eggl moved to Paris, France, in 2013 after completing his PhD thesis on the subject of the habitability of terrestrial planets in binary star systems at the Institute for Astrophysics of the University of Vienna, Austria. There, he worked as a postdoctoral research associate at the *Institut de Mécanique Céleste et de Calcul des Ephémérides* (IMCCE) in planetary defense. In 2016, Dr. Eggl accepted a research appointment at the NASA Jet Propulsion Laboratory, California Institute of Technology in Pasadena, California, USA.

Ákos Bazsó graduated with a PhD in 2015, from the University of Vienna, Austria. Since then, he's worked there as a postdoc researcher, specialising on extending the concept of habitability to binary star systems. His main fields of research are the dynamics of planets and minor bodies in the solar system, as well as dynamics and resonances in extra-solar planetary systems, including binary and multiple star systems.

Acknowledgments

The authors acknowledge the long years of support by the Austrian Science Fund (FWF) through projects: P20216-N16 "Evolution of planetary systems in binaries" (*EP-L & SE*), P22603-N16 "Exoplanetary systems: Architecture, evolution, habitability" (*EP-L & AB*), and NFN sub-project S11608-N16 "Binary Stars and Habitability" (*EP-L, SE & AB*), which is part of the FWF/NFN project: S116 "Pathways to Habitability: From Disks to Active Stars, Planets to Life" (headed by Prof. Manuel Güdel).

SE also received funding for this research from the Jet Propulsion Laboratory through the California Institute of Technology postdoctoral fellowship program, under a contract with the National Aeronautics and Space Administration, USA, and from the IMCCE Observatoire de Paris, France.

We wish to thank our colleagues from the Department of Astrophysics at the University of Vienna for their collaborations in the field of planetary dynamics and habitability in the framework of the above mentioned projects. In particular, discussions with Dr. David Bancelin on the water transport via planetesimals into the habitable zone, and Mag. Markus Gyergyovits on the early stages of planetary formation were very fruitful for this monograph. Dr. Barbara Funk together with Dr. Martin Netopil contributed information on the various S- and P-type systems in Chapter 7. Moreover, Dr. Thomas I. Maindl provided valuable SPH simulations of planetesimal collisions with protoplanets, and Dr. Barbara Funk contributed the sketch designs in Chapters 2 and 5.

Furthermore, we want to thank to our international collaborators: Prof. Nader Haghighipour for a long lasting cooperation on binary stars and for the invitation of two authors (*EP-L & SE*) to the Institute for Astronomy of the University of Hawaii in the summer of 2012; Dr. Philippe Robutel from IMCCE, Paris (France) for numerous discussions about resonances in planetary systems which were very fruitful for our research; Dr. Nikolaos Georgakarakos from the New York University (Abu Dhabi) whose studies were applied for the analytical approach to calculate the habitable zones in binary star systems; Prof. Zsolt Sándor from the Eötvös Loránd University in Budapest (Hungary) for providing the figures about terrestrial planet formation in Chapter 4.

Our special thanks go to Doz. Dr. Helmut Lammer who encouraged us to perform habitability studies from the dynamical point of view and who is always very enthusiastic about collaborations, and Dr. Colin Johnstone for the proofreading of the monograph.

We cordially thank our supervisor Prof. Rudolf Dvorak who founded the Astro Dynamics Group (ADG) at the University of Vienna and introduced us to the fascinating science of planetary dynamics. Last but not least, we want to thank Prof. Wing-Huen Ip and the World Scientific Publishing Company for the invitation to give a review of our research from the past years in the book series "Advances in Planetary Sciences".

Contents

Background

Astrophysical research has led to the detection of thousands of planets outside the Solar System. About one tenth of the extrasolar planets discovered so far reside in binary or multi-stellar systems, and some of the closest known rocky exoplanets populate these multiple star systems. While such environments seem good places to look for a second Earth, can Earth-like planets with two or more suns be habitable? And do solar system-like configurations have to be detected to find a habitable exo-Earth?

This book addresses these questions. Starting with a brief overview of the various types of double star–planet configurations that have been observed so far, the book discusses the intriguing variety of planetary motion in such environments, taking into account the stellar type (its evolution and activity) and elaborates on how the presence of an additional stellar companion affects planet formation, system architectures and the habitability of planets in binary star systems. New methodologies developed in this area of research are explained and demonstrated for systems such as Alpha-Centauri, HD41004, Kepler-35, and many others. This monograph provides a grand entry to the exciting results that we expect from new missions like TESS, CHEOPS and Plato.

Contents: Binary Stars; Orbital motion in Binary Star Systems; Perturbations in Multi-Planet Binary Star Systems; Terrestrial Planet Formation in Binary Stars; Implications of Stellar Binarity; Habitable Zones in Binary Star Systems; Habitability of known planets in Binary Star Systems.

Chapter 1

Binary Stars

In astrophysics, binary stars are very useful, because they provide the best method to directly determine the masses of distant stars since they orbit around their common center of mass. A large fraction of stars in the solar neighborhood form binary or multiple star systems, thus they are important for our understanding of processes by which stars and planets form. Therefore, we review in this chapter some basics about binary stars and issues relating to habitability in binary star systems.

1.1 Classification

Sir William Herschel introduced the term *binary stars* in 1802 when he cat-alogued his observations of approximately 700 double-stars (Heintz, 1978). Further discoveries of such systems led to the following classification:

Visual binaries: These are two stars that have a wide enough orbital separation to be viewed separately with a telescope. The first observed binary stars, e.g., *Mizar* (ζ UMa), belong to this category.

Spectroscopic binaries: These are two stars that have a small separation even when they are observed through a telescope. It is difficult to spatially resolve the two components. They can only be detected in the combined spectrum of the system using the Doppler shift of spectral lines due their orbital motion.

Eclipsing binaries: These are two close stars that cause an eclipse when one passes in front of the other while observing them. From the periodic changes in brightness, one can determine easily the orbital period. Also, two distant stars can show eclipses if the viewing geometry is favourable, but typically the probability for eclipses is higher when the orbital separation is smaller.

Astrometric binaries: Observations indicate that they are single stars because their companions cannot be seen; maybe they are too dim or they are hidden because the primary star is so much brighter. Therefore, one can only indirectly infer that a secondary star is present from the orbital motion of the primary around the center of mass of the system.

For dynamical studies, the orbital separation of the two stars is the key parameter, leading to a classification into wide and close binaries. The orbital periods can be anything from less than an hour to hundreds of thousands of years. Our closest neighboring star belongs to a triple system consisting of a close binary (α Centauri AB) and a distant M-star (Proxima Centauri), which is currently of special interest due to the recent detection of a planet in the habitable zone.

Wide binaries have orbital separations of several thousands of astronomical units (au) and evolve independently with very little impact from their companions so that they can be considered as single stars. This was the common belief until the recent study by Kaib *et al.* (2013). They figured out that a faraway companion star could have significant changes in its orbital motion due to the influence of the galactic tide or impulses from a passing star. Such changes could perturb an otherwise stable system of planets orbiting the other stellar component, leading to ejections of the outer planets and exciting the orbital eccentricities of the remaining bodies. Such a scenario could also have severe consequences for the habitability of a planetary system.

Close binaries are characterized by a significant gravitational influence of their companion stars. These systems are of high interest for dynamical studies. If the separation is only a few stellar radii then the gravitational interaction could strongly influence the stability of the bodies, in extreme cases leading to the destruction of a body. For close binary stars, the physics of the stellar system can be described via the Roche lobe which defines a critical sphere around each star. When the two stars are close enough, their Roche lobes might be connected at the Lagrangian equilibrium point L_1. Hence, different binary configurations can be distinguished:

Detached binaries are where each star is inside its Roche lobe without significant gravitational effects on the stellar structure by the other stellar component, so the two stars evolve separately.

Semi-detached binaries: When one component fills the Roche lobe while the other does not. In this case, there can be a mass transfer from the Roche-lobe filling component to the other, which is important for the evolution of the binary system.

Contact binaries: When both stars fill their Roche lobes so that a common envelope can be formed, which surrounds both stars. Depending on the mass and the type of the two objects, such a binary could lead to a merging of both bodies.

1.2 Observations

Numerical simulations of the collapse of a rotating cloud by Larson (1972) show that preferentially binary and multiple star systems are formed, while single stars are rather escapers from dynamically unstable multiple systems. The mixture of single, binary and multiple stars at different ages and in different environments indicates that these systems form most probably due to dynamical interactions as part of the early stellar evolution. Binary stars are the most common multiple star systems where two stars are orbiting a common center of mass. The brighter star is usually classified as the primary star S_A, while the dimmer star is the secondary star S_B.

Similarly, observations suggest that a considerable fraction of stars in the solar neighborhood are members of binary and multiple star systems. Duquennoy *et al.* (1991) and Raghavan *et al.* (2010) established that in the solar neighborhood ($d < 25$ parsec), about 40–45% of all Sun-like stars (spectral types F6–K3) are members of binary and multiple star systems, independent of whether or not they are hosting planets. More recently, Tokovinin (2014) found for a sample of about 4,800 F-/G-type main-sequence stars within 67 parsec of the Sun, that 33% of the targets belong to binary star systems. For a more detailed study on this topic, we refer the reader to the review of Duchêne and Kraus (2013).

Recently, Schwarz *et al.* (2016) summarized the data published about binary star systems from about 20 catalogues, where each catalogue provides information on either a few dozen or up to some thousands of systems. The largest catalogue is the *Washington Double Star Catalogue*[1] (WDS) which is a collection of more than 130,000 binaries. The key parameters for dynamical studies of planetary motion in binary star system are the separation of the two stars, i.e., the binary's semi-major axis (a_B) and the eccentricity (e_B) of their orbits. The fact that a_B and e_B are known only for a small fraction of observed binary star systems indicates the difficulty for dynamical studies due to the lack of observational information.

In Figure 1.1, one can see the distribution of a_B and e_B for about 660 binary star systems with well determined parameters published in the WDS.

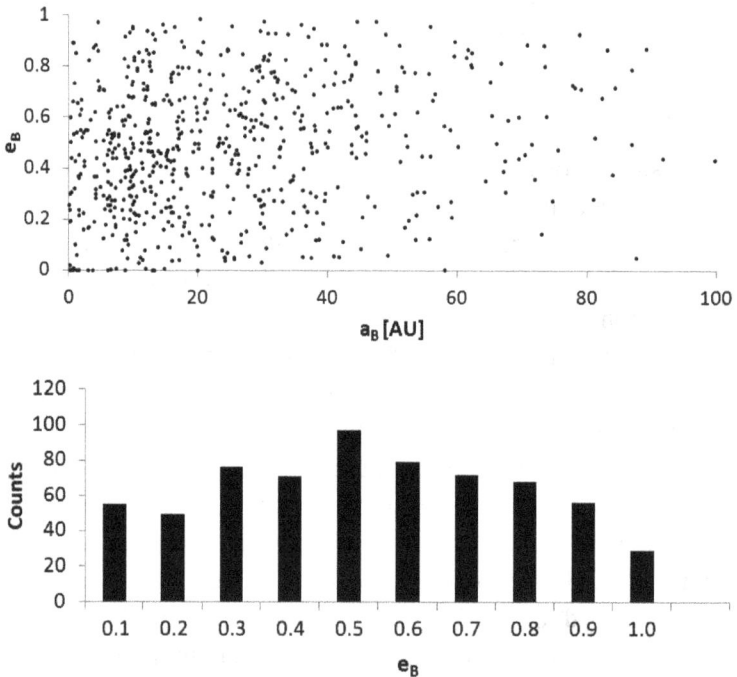

Figure 1.1: Top panel: Distribution of binary semi-major axis (a_B) and eccentricity (e_B) for systems from the Washington Double Star Catalogue. Bottom panel: Bar chart of the binary's eccentricity.

---

[1] http://ad.usno.navy.mil/wds/.

The top panel of this figure shows clearly that eccentricity is uncorrelated with a_B. The only exception to this lack of correlation is for extremely tight binaries ($a_B < 0.1$ au), which generally have no eccentricity due to tidal interactions between the two stars (Halbwachs *et al.*, 2003). The lower density of data points for larger a_B reflects the lack of information (or observations) of wide binary systems. In the bottom panel of Figure 1.1, we show the distribution of the eccentricity of the 660 binary stars. The plot indicates a slight peak at $e_B = 0.5$. It should be noted that the information of a_B and e_B cannot be provided without large errors from the observations. This restricts also the accuracy of dynamical studies.

1.3 Planets in Binary Star Systems

The detection of extra-solar planets in binary and multi-stellar systems showed that planetary companions are not restricted to single stars. However, we cannot conclude whether these environments are more hostile for the presence of planets or not (see e.g., Boss, 2006; Bromley and Kenyon, 2015; Jang-Condell, 2015). A study by Armstrong *et al.* (2014) using the *Kepler* data suggests an occurrence rate of coplanar circumbinary planets similar to that for single stars. For habitability studies, we are mainly interested in host-stars of spectral types F, G, K and M because the lifetimes of these stars on the main sequence are sufficiently long to permit the evolution of life (as known on Earth) on an Earth-like planet in the habitable zone (see Kasting *et al.*, 1993).

In the top panel of Figure 1.2, we show the number of different binary pairs for these spectral types published in the WDS catalogue. For this figure, we have gathered the information of more than 7,000 systems. The significant peak for M–M pairs indicates the predominance of M-type stars in the solar neighborhood. Moreover, the distribution shows a slightly higher occurrence rate for stars with similar spectral types than for two stars of different types, i.e., F–M pairs have the lowest frequency. In the same manner, in the bottom panel of Figure 1.2, we show a distribution of all binary star systems that host one or more planets , which is based on the data in the 'Binary catalogue' of Schwarz *et al.* (2016). This histogram indicates that binaries with M-type secondaries are more frequent than other stellar types — an exception is the significant peak for G–K binary pairs. Of course, the available sample is still small and such preferences of

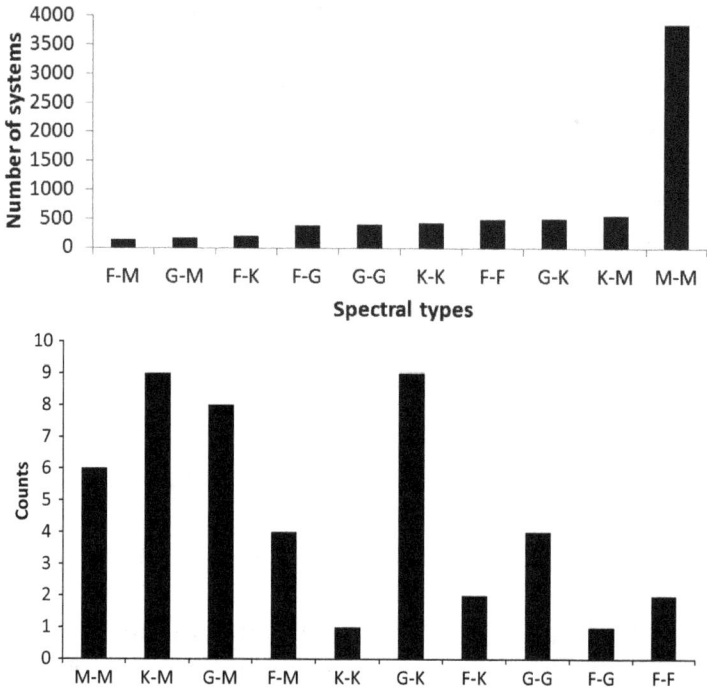

Figure 1.2: Top panel: Distribution of spectral types of binary pairs, restricted to stars with spectral types F to M from the WDS catalogue. Bottom panel: Distribution of spectral types for planet-hosting binaries.

star combinations could disappear when the number of observed systems is significantly increased.

The top panel of Figure 1.3 shows the binary orbital separations (a_B) as a function of either the primary star's mass (for S-type planets) or of the stellar mass-ratio (for P-type planets) for binary systems that contain planets. Different types of planetary motion[2] are pictured by different colors (as labeled in the figure). One can recognize clearly two distinct areas where the upper cloud is populated by S-type planets and the lower one by P-type planets. A single exception is the blue dot at $a_B \sim 10$ au, which denotes a planet in P-type motion in the S-type area (FW Tau AB b, see Table 7.2 in Section 7.1). In fact, we expect that further observations will reveal a lower

---

[2]For definition see Chapter 2.

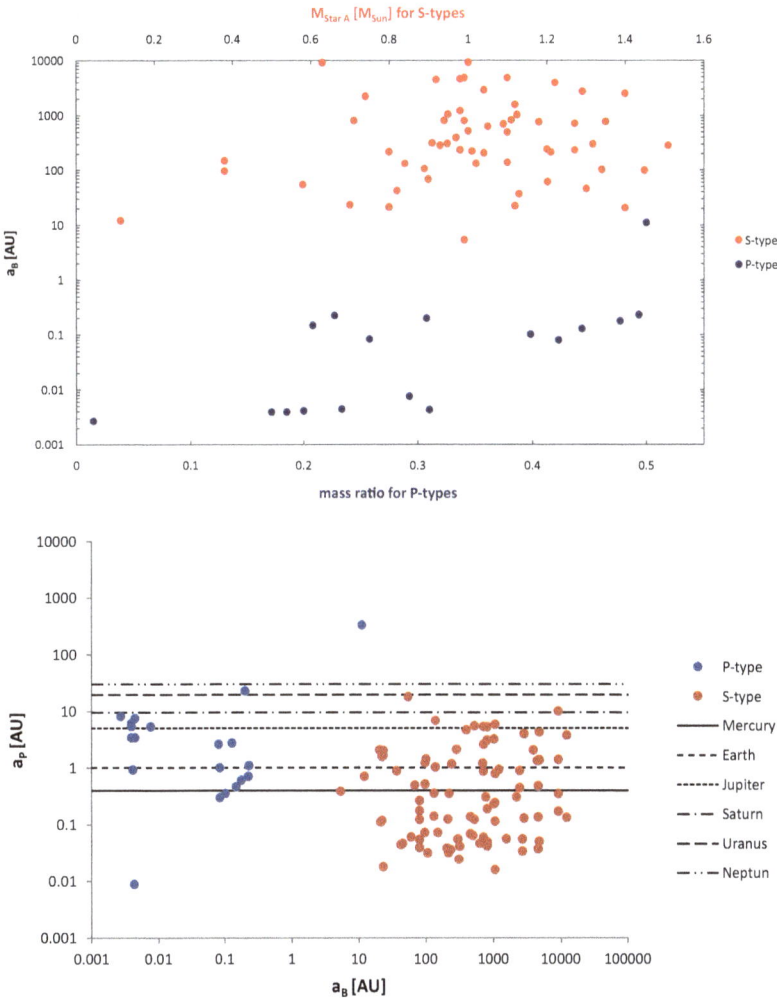

Figure 1.3: Top panel: The semi-major axis a_B of the secondary star as a function of the primary star's mass for S-type and of the star's mass-ratio for P-type planets. Bottom panel: Semi-major axes of S-type and P-type planets as a function of the secondary star's semi-major axis and a comparison with several planets of the Solar System (labeled by the various horizontal lines).

limit for the distance of the two stars, in case of S-type motion which at present is around 5 au according to observations. For P-type planets, it can be expected that future detections will populate the empty region between the two clouds of data points.

The separation of P-type and S-type planets is also visible in the bottom panel of Figure 1.3, which shows all detected planets in binary star systems in a plane where we plotted the stellar separations versus the planet's distance to its host-star. In addition, we compare the planet's orbital distance, a_P, with those of some Solar System planets indicated by the different horizontal lines. In general, P-type planets (blue dots) are found at larger orbital distances and most of these planets have semi-major axes larger than that of Mercury (indicated by the solid horizontal line). This is certainly for stability reasons (see Chapter 2). Surprisingly, the case with the smallest planetary orbital distance is a P-type system and is the only notable exception to the trend mentioned above (i.e., the blue dot in the lower left corner). In contrast, the cloud of S-type planets (red dots) show a variety of distances that are all closer than of Uranus in our Solar System, which is shown by the dashed horizontal line. Of course, we expect more distant planets also in S-type motion, especially in wide binary star systems with stellar separations of several tens of thousands of astronomical units. The plots shown in Figure 1.3 map the current state of observations which will certainly change a lot in the future.

In any case, multi-body systems need to be considered carefully since the gravitational interactions of the bodies restrict the stability for orbital motion to certain regions of the phase space. This underlines the necessity of dynamical studies because the long-term stability of a planetary system is a basic requirement for habitability.

Chapter 2

Orbital Motion in Binary Star Systems

In dynamical astronomy, planetary orbits are usually described by a set of six orbital elements. The size and the shape of the orbit are determined by the semi-major axis a and the eccentricity e. The orientation in space is determined by three angles: the inclination i, the argument of perihelion ω, and the longitude of the ascending node Ω. The mean anomaly, M, gives the position of the celestial body in the orbit. A graphical illustration of these parameters is given in Figure 2.1.

Five of the six orbital elements are constants of motion resulting from the solution of the two-body problem (e.g., a planet orbiting a star), which is the simplest model in planetary dynamics and the only problem with a fully analytic solution. Usually the two bodies are considered as point masses moving under the mutual gravitational attraction according to Newton's universal law of gravitation. A planet's motion around a star is described by the equation of relative motion in the heliocentric coordinate frame.

For planetary motion in binary star systems, we have to consider a three-body model where the planet moves in the gravitational field of two stars, which requires a barycentric (center of mass) coordinate system. The equations of motion are given by

$$m_i \frac{\mathrm{d}^2 \vec{r}_i}{\mathrm{d}t^2} = \mathcal{G} m_i \sum_{j \neq i} m_j \frac{\vec{r}_{ji}}{r_{ji}^3}, \qquad (2.1)$$

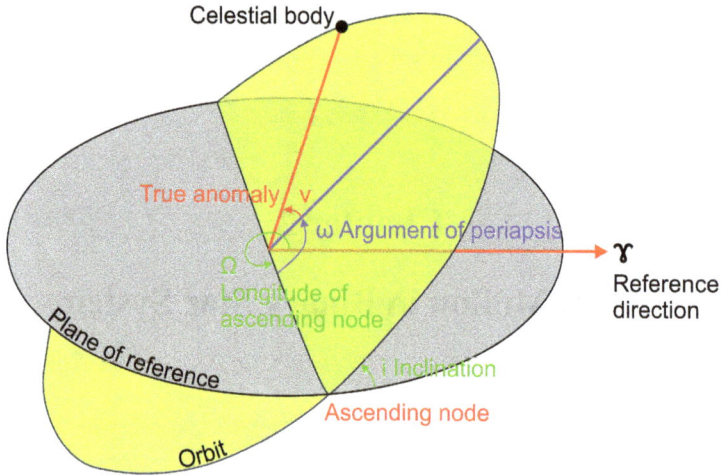

Figure 2.1: Illustration of the orbital elements (i, ω, Ω and υ) describing the orientation of the orbit in space.

for i, $j = 0, 1, 2$, with the relative position vector $\vec{r}_{ji} = \vec{r}_j - \vec{r}_i$; $r_{ji} = \|\vec{r}_{ji}\|$. The variables m_0 and m_1 are the masses of the two stars, m_2 is the mass of the planet, and $\mathcal{G} = 6.6726 \times 10^{-11}$ m^3 kg^{-1} s^{-2} is the universal gravitational constant. For a fully analytical solution, 18 constants of motion are needed, but only 10 are known, so numerical solutions are required for the long-term orbital evolution in the three-body problem.

When one body is considerably less massive than the other two, one can use the *restricted three-body problem* (RTBP). For studies of planetary motion in binary star systems we can apply the RTBP and use that the mass of the planet is negligible relative to the masses of the two stars. The planetary motion is perturbed by the two stars, which themselves move on Keplerian orbits without being perturbed gravitationally by the planet. For details about celestial mechanics, we refer the reader to Roy (1988), Murray and Dermott (1999) and other textbooks.

The RTBP has been used for general stability studies of planetary motion in binary star systems that have been carried out long before the detection of planets outside the Solar System (see Harrington, 1977; Szebehely, 1980; Szebehely and McKenzie, 1981; Dvorak, 1984, 1986; Rabl and Dvorak, 1988; Dvorak *et al.*, 1989). Also, several studies of real binary systems have been carried out using the RTBP (see Benest,

1988a,b, 1989, 1993, 1996, 1998). The various investigations showed clearly that stable planetary motion in binary systems is restricted to certain regions in the phase space which depend on the binary–planet configuration. Moreover, in a binary star system one has to consider different types of planetary motion.

2.1 Types of Planetary Motion

From the dynamical point of view, we can distinguish the following types of motion in double-star systems (Dvorak, 1984):

(1) **P-type or circumbinary motion:** The planet orbits both stars in a distant orbit that has to be inside the grey area in the top panel of Figure 2.2, which is outside the critical distance a_{crit} as indicated by the red circle.

(2) **S-type or circumstellar motion:** The planet orbits one star as shown in the middle panel of Figure 2.2; this can be either the primary S_A or the secondary S_B. The planetary orbit has to be close enough to the host-star, inside the grey area in the middle panel of Figure 2.2, which is delimited by the critical distance a_{crit} (red circle). Observations have revealed that there are also binary star systems where both stars have substellar companions (e.g., HD 41004 AB).

(3) **T-type or librational motion:** The planet moves in the same orbit as the secondary star but $60°$ ahead or behind (see the grey area in the sketch in the bottom panel of Figure 2.2). The two bodies are locked in 1:1 mean motion resonance to stabilize their motion. These areas of stable planetary motion are extending around the two triangular Lagrangian points in the orbit of the secondary star.

However, the T-type motion is not that important for binary star systems, because there is a limitation on the mass-ratio of the two stars ($\mu = m_2/(m_1 + m_2)$) which requires $\mu < 1/26$. A study by Schwarz *et al.* (2015) found some examples of binary systems that fulfill this condition. It is obvious that one stellar component has to be very massive (O or B type star), while the other component must be a low mass star. This leads to the conclusion that this type of motion is less interesting for habitability studies in binary star systems. T-type motion is more applicable for single

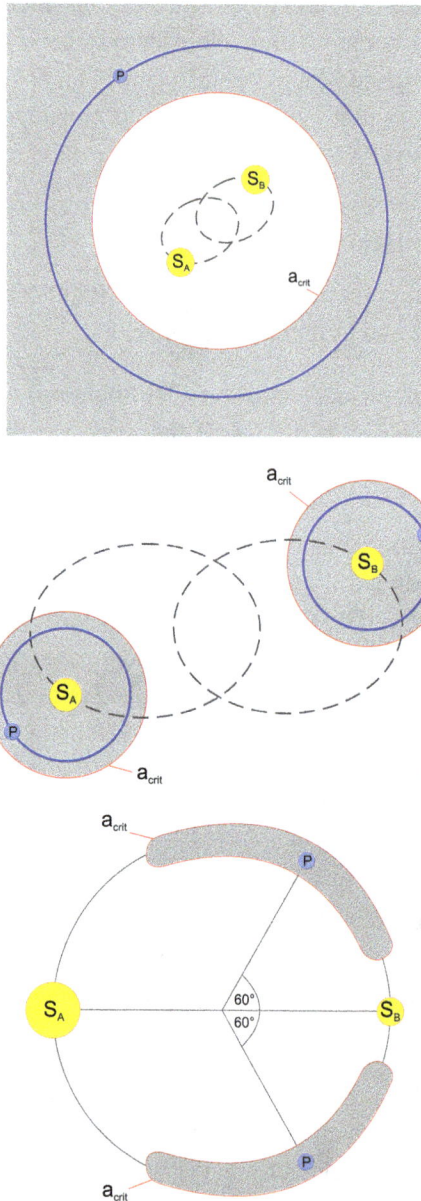

Figure 2.2: Different types of planetary motion in binary star systems. Top: P-type; middle: S-type; bottom: T-type motion. The shaded areas indicate stable motion and the red orbits label the stability limits that are defined by the critical distance a_{crit}.

star–giant planet configurations which easily fulfill the condition above, and it permits more than one planet to move in an orbit inside the HZ. In this book we concentrate on S-type and P-type planetary motion since all detected planets in binary systems belong to one of these two types.

Observations of extra-solar planets in binary and multiple-stellar systems have shown that planetary companions are not restricted to single-stars. In addition, these discoveries encouraged various research groups to reproduce the general stability studies (see e.g., Holman and Wiegert, 1999; Pilat-Lohinger and Dvorak, 2002; Pilat-Lohinger *et al.*, 2003) which are also an important starting point for habitability studies. Long-term stability is a basic requirement for the habitability of an Earth-like planet, which can be concluded from the evolution of the biosphere on Earth.

2.2 Stability of Circular Planetary Orbits

The fact that the planets in our Solar System move in nearly circular orbits aligned nearly in the same plane had certainly an effect on the stability studies carried out either long before the detection of extra-solar planets or at the beginning of the exoplanetary research era. In the following, we review briefly the stability of circular and eccentric planetary motion and show the application for habitability studies from the dynamical point of view. Using the RTBP (since the planetary mass is much smaller than the stellar masses), the stability for S-type and P-type motion in binary star systems is determined as follows:

(1) The distance between the stars is defined as the unit distance.
(2) The semi-major axis of the planet (a_P) is varied to determine the critical value a_{crit}, which gives the stability border.
(3) The mean anomaly of the planet, M_P, is varied for each a_P to ensure that the stability border is independent of the planet's position.

2.2.1 *S-type motion*

For circumstellar/S-type motion, a_{crit} is the outermost distance from the host-star up to which planetary motion is stable independent of the initial position of the planet in its orbit. The investigation by Rabl and Dvorak (1988) was the first that determined the stability limit as described above for the mass-ratio $\mu = 0.5$ and various binary eccentricities between 0 and 0.6. By defining lower and upper critical orbits (LCO and UCO, respectively), they showed that between these orbits there exist a "grey zone" where both regular and chaotic motion may occur for a certain distance of the planet. This depends on the initial position of the planet which is defined by the variation of the mean anomaly M_P. The red curve in the middle pannel of Figure 2.2 defines the outer border of stable motion and corresponds to the LCO defined by Rabl and Dvorak (1988).

A decade later, when the first exoplanets in tight binary star systems have been detected (e.g., γ Cephei, Gliese 86), Holman and Wiegert (1999) repeated this study, taking into account various mass-ratios, μ, from 0.1 to 0.9, and eccentricities, e_B, from 0.0 to 0.8, and a longer computation time of 10,000 periods of the binary. The main result of this study was an empirical relation for how a_{crit} depends on μ and e_B:

$$a_{\text{crit}} = \big[(0.464 \pm 0.006) + (-0.380 \pm 0.010)\mu$$
$$+(-0.631 \pm 0.034)e + (0.586 \pm 0.061)\mu e$$
$$+(0.150 \pm 0.041)e^2 + (-0.198 \pm 0.074)\mu e^2\big]a_B. \quad (2.2)$$

Holman and Wiegert (1999) showed that their results for $\mu = 0.5$ are in good agreement with those of Rabl and Dvorak (1988). However, since both of these studies used only straightforward computations without analyzing the orbits, Pilat-Lohinger and Dvorak (2002) repeated the study using the Fast Lyapunov Indicator (FLI) to distinguish between stable and chaotic motion. The FLI is a well known chaos indicator introduced by Froeschlé *et al.* (1997) that determines chaos via the growth of the largest tangent vector of the trajectory. A linear growth of this vector denotes regular planetary motion while an exponential growth indicates chaos for the orbit. A comparison of the FLI study with the results of Holman and Wiegert (1999) and those of Rabl and Dvorak (1988) shows a deviation of about 2% in the extension of the stable region, which proves that the different studies are in good agreement.

In Figure 2.3, the stable zone around one stellar component is summarized for mass-ratios, μ, from 0.1 to 0.9 of the binary (see the x-axis) and from circular binary motion up to $e_B = 0.8$ (y-axis). One can recognize mainly a linear decrease of the stable area, defined by a_{crit},

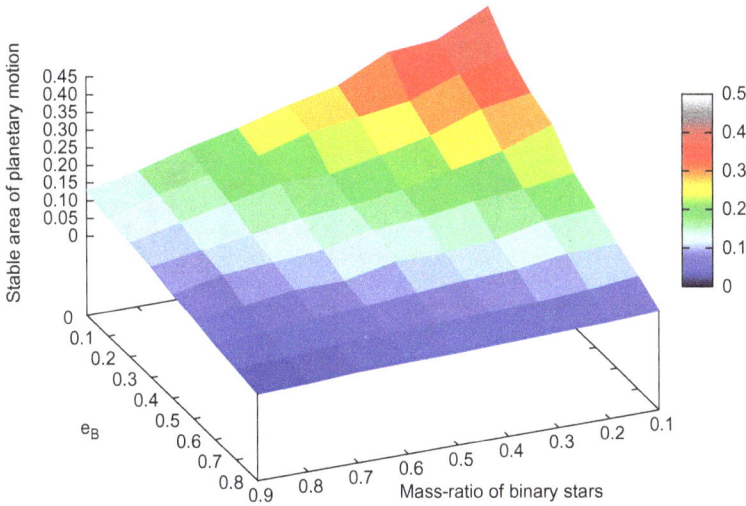

Figure 2.3: Overview of the stable zone for circular planetary S-type motion in various binary systems with mass-ratios between 0.1 and 0.9 and stellar eccentricities between 0 and 0.8. The colors indicate the size of the stable area around the host-star. For details, see the text.

on the z-axis when increasing e_B. The color scale shows the size of the stable area (in dimensionless units) of S-type motion. This is always less than half the distance between the two stars, defined by a_B, even if the mass of the perturbing star is small ($\mu = 0.1$) compared to the mass of the host-star. This pair of masses yields the largest stable area around m_A, with $a_{\text{crit}} = 0.45$ as indicated by the red zone. The color code in Figure 2.3 shows clearly that only for low mass-ratios and low eccentricities of the binary (e_B) is the stable area larger than a quarter of the mutual stellar distance a_B. Most of these binary configurations have a stable region of less than one-fifth of a_B (see the green and blue areas).

These results from Pilat-Lohinger and Dvorak (2002) were used to determine the minimum distance of the two stars for different binary configurations when requiring that the HZ belongs to the stable area. Therefore, the outer border of the HZ was calculated according to Kopparapu *et al.* (2014) for the various spectral types (F to M) that are interesting for habitability studies. The resulting HZ borders are shown in Table 2.1.

Table 2.1: HZ borders for different main-sequence stars.

Spectral type	Mass $M/M_\odot$	Luminosity $L/L_\odot$	HZ borders [au] Inner	HZ borders [au] Outer
F	1.30	2.500	1.436	2.493
G	1.00	1.000	0.950	1.676
K	0.69	0.160	0.406	0.755
M	0.47	0.063	0.258	0.455

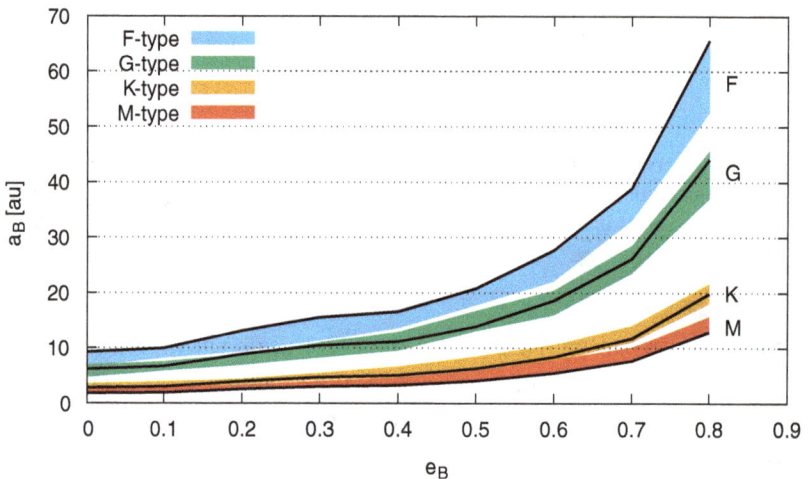

Figure 2.4: Minimum distance of the two stars as a function of the binary's eccentricity such that the HZ is entirely in the stable area shown for different binary configurations. The different colors indicate different stellar types for the host-star and the width of the colored area shows the minimum distance for different stellar types of the secondary star (from F- to M-type). Black lines indicate the minimum distances for equal-mass binary configurations.

From recent calculations of the circumstellar HZ in binary star systems (see Eggl *et al.*, 2012, 2013a,b), we can assume that these values correspond fairly well to the averaged HZ (AHZ, see Chapter 6) of the binary star configurations. For S-type motion, the dynamical stability of the HZ is satisfied if a_{crit} is larger than the outer border of the HZ, which depends on the stellar type, as well as the semi-major axis, a_B, and eccentricity, e_B, of the binary. Figure 2.4 shows the minimum distance of the two stars as a function of the binary's eccentricities, where the different colors label different stellar types for the host-star. The width of each colored band results from the minimum distance for various stellar types of the secondary star, where the upper border corresponds to the results of an F-type secondary and the lower border to those of an M-type companion. The black line in a colored band indicates the minimum distance when the two stars have the same mass. This figure shows clearly that low-mass stars (M- and K-types) provide dynamical habitability even in tight binary configurations (like γ Cephei with $a_B \sim 20$ au) with high eccentricities (up to 0.8). For a more massive host-star in such tight configurations, the eccentricity needs to be less than 0.6 for a G-type star, or less than 0.5 for an F-type star. For larger eccentricities, the G- and F-type host-stars indicate an exponential increase of the distance of the two stars (a_B) to fulfill the requirement of a stable HZ.

2.2.2 P-type motion

For P-type planetary orbits, in which the planet orbits both stars, an empirical rule states that the planet's semi-major axis measured from the center of mass of the two stars has to be at least twice the mutual distance of the two stars (strictly speaking, this applies for $\mu = 0.1$ and $e_B = 0$). It is quite obvious that the value of a_{crit} will increase (i.e., the stability border is shifted away from the two stars) when the mass-ratio μ and/or the eccentricity e_B is increased. Moreover, former stability studies (see e.g., Dvorak *et al.*, 1989; Holman and Wiegert, 1999) of circular planetary motion in P-type systems showed that the influence of the binary's eccentricity is stronger than that of the mass-ratio.

As an example, we show in Figure 2.5 the stability border of P-type motion around two G-type stars for circular and eccentric motion of the binary ($0 \leq e_B \leq 0.5$). The different colors indicate different maximum eccentricities of test-planets that are orbiting the binary between 1.8 and 4.5 length units.[1] Blue areas show stable motion, red the unstable region, green and yellow color show an intermediate zone, i.e., the 'grey zone' according to Dvorak (1986), where both stable and unstable motion has

---

[1] Since we define the separation of the two stars as unit distance, a_{crit} is dimensionless.

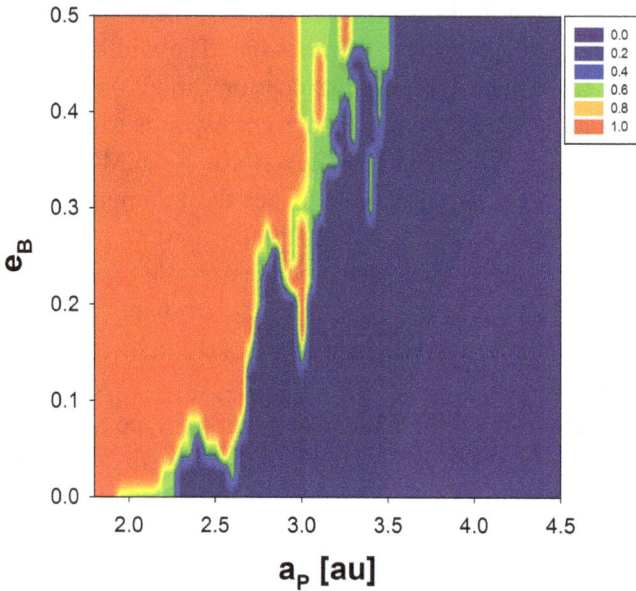

Figure 2.5: Stability of circular planetary circumbinary motion around two G-type stars for different eccentricities of the binary. The colors indicate different maximum eccentricities of the planetary motion.

been found for a certain distance of the test-planet depending on its initial position. Figure 2.5 shows clearly that the stability border is not given by a smooth curve because of mean motion resonances with the binary star system, which obviously play an important role. For binary eccentricities less than 0.3, we recognize that the borders of stable and unstable motion are close to each other so that the intermediate zone is very small. This changes for higher e_B where a large green area is visible. According to the blue region, we identify the stability border a_{crit} at $a_P = 2.3$ for circular binary motion and at about $a_P = 3.6$ for $e_B = 0.5$. These results are in good agreement with those of Holman and Wiegert (1999), which led to the following expression for a_{crit} of P-type motion:

$$a_{crit} = \big[(1.60 \pm 0.04) + (5.10 \pm 0.05)e + (-2.22 \pm 0.11)e^2$$
$$+ (4.12 \pm 0.09)\mu + (-4.27 \pm 0.17)e\mu + (-5.09 \pm 0.11)\mu^2$$
$$+ (4.61 \pm 0.36)e^2\mu^2\big]a_B. \tag{2.3}$$

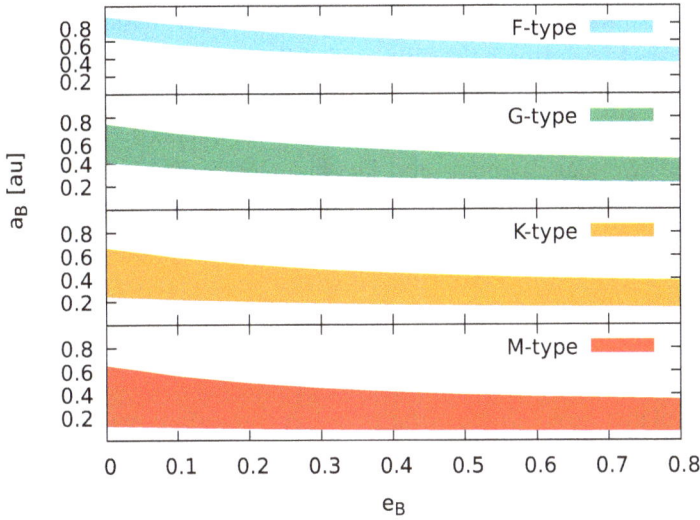

Figure 2.6: Maximum distance of the two stars as a function of the binary's eccentricity such that the HZ is entirely in the stable area shown for different binary configurations. The different colors indicate different stellar types: Blue for binary stars with at least an F-type star, green with an G-type star, orange with an K-type star and red with an M-type star.

A necessary condition for a habitable P-type planet is the long-term dynamical stability of planets in the HZ. Therefore, the HZ has to lie in the stable area. To fulfill this condition, the semi-major axis of the inner HZ border has to be larger than a_{crit}. Taking into account the general stability studies and the HZ borders (see Table 2.1), we can conclude that the distance between the two stars should be less than 1 au for circular binary motion, or less than 0.5 au for highly eccentric motion of the two stars, to grant long-term dynamical stability for P-type planetary motion in the HZ (as shown in Figure 2.6).

2.3 Stability of Eccentric Planetary Orbits

In contrast to the nearly circular planetary motion in our Solar System, we learned from observations of exo-planets that planetary motion can be highly eccentric. The extrasolar planet database[2] indicates eccentricities up

---

[2]http://exoplanet.eu.

to 0.6 for planetary masses smaller than 0.05 Jupiter-masses. Therefore, it is important to determine the influence of the planetary eccentricity on the extension of the stable area.

The first study that took into account the planet's eccentricity was published by Pilat-Lohinger and Dvorak (2002). They showed that the binary's eccentricity has a stronger effect on the extension of the stable area than the planet's eccentricity. However, Pilat-Lohinger (2012) pointed out that the result strongly depends on the mass of the planet.

2.3.1 *S-type motion*

When a planet orbits its host-star in an eccentric orbit caused by gravitational perturbations, either of the secondary star (see Chapter 5) or of an additional planet, the stable circumstellar area shrinks. How much it shrinks depends on the stellar parameters (the mass-ratio, the semi-major axis and the eccentricity). These changes are visualized in Figure 2.7, where a_{crit} is shown for several values of e_B from Figure 2.3.

In Figure 2.7, we show the changes in the extension of the stable area for three binary eccentricities ($e_B = 0.1$, 0.3 and 0.5) and all mass-ratios from 0.1 to 0.9 (y-axis). An increase of the planet's eccentricity is shown on the

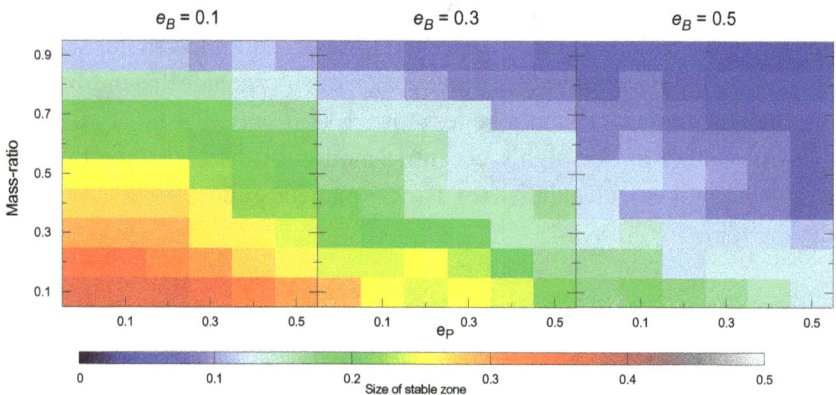

Figure 2.7: Stability of eccentric planetary S-type motion in binary star systems for different eccentricities of the binary.

x-axis. The results for circular planetary motion (i.e., $e_P = 0$) are located in the leftmost position of each panel. Comparing the colors of the three panels, one recognizes the strong shift of a_{crit} towards the host-star when increasing the eccentricity of the binary e_B (labels at the top). Furthermore, for a low binary eccentricity (e.g., $e_B = 0.1$), an increase of e_P does not change a_{crit} for a certain mass-ratio. We only see minor variations for planetary eccentricities larger than 0.3. A moderate binary eccentricity ($e_B = 0.3$) indicates more changes especially for lower mass-ratios of the binary (0.1 and 0.2). For large binary eccentricities (right panel), the stable area around the host star shrinks for most configurations to less than about 10% of the stellar distance. However, Figure 2.7 shows clearly the stronger influence of the binary's eccentricity as was already pointed out in Pilat-Lohinger and Dvorak (2002).

A similar study for a massive planet has been carried out to analyze the influence of the planetary mass on the result and to gather information about the limits of validity of the RTBP. For a fixed[3] mass-ratio of $\mu = 0.2$,

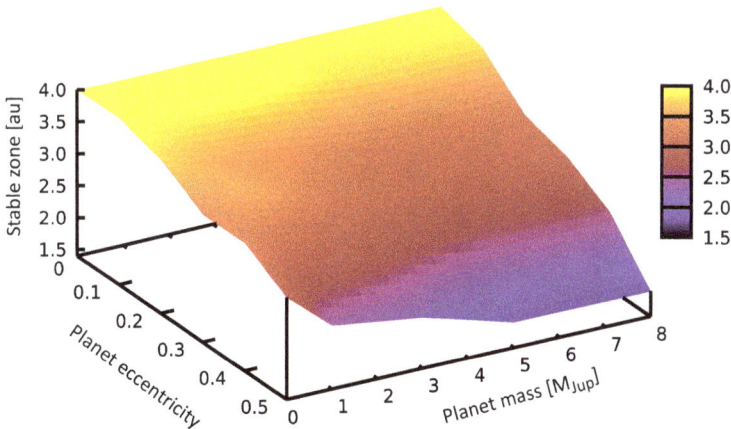

Figure 2.8: Border of stable region for massive planets in eccentric circumstellar (S-type) motion.

---

[3]The mass-ratio 0.2 was chosen to apply this study to the real binary system γ Cephei.

the planetary mass, m_P, was varied from 0 to 8 Jupiter-mass, and its eccentricity, e_P, from 0 to 0.5. For this set of parameters the stability border of the various binary–planet configurations was calculated. Such systems represent γ Cephei-like systems (see Chapter 7 for the parameters), where the planet orbits the more massive star and the perturbing secondary star is at a distance of about 20 au. Taking into account the binary's eccentricity, $e_B = 0.4$, the border of the stable region for the different configurations are shown in Figure 2.8, which indicates clearly that when the planetary motion is circular, the planetary mass has no visible influence on the result (see the yellow area for $e_P = 0$ and all planetary masses, m_P). The result of the RTBP (i.e., $m_P = 0$) indicates a shift of the stable border from 4 au to 3 au when the planetary eccentricity is increased from 0 to 0.5. A stronger shift of the stable border towards the host-star can be recognized especially for higher planetary masses (5 or 8 Jupiter-masses) in highly eccentric motion, where the radius of the stable zone shrinks from 4 au to 1.8 au. Thus, the

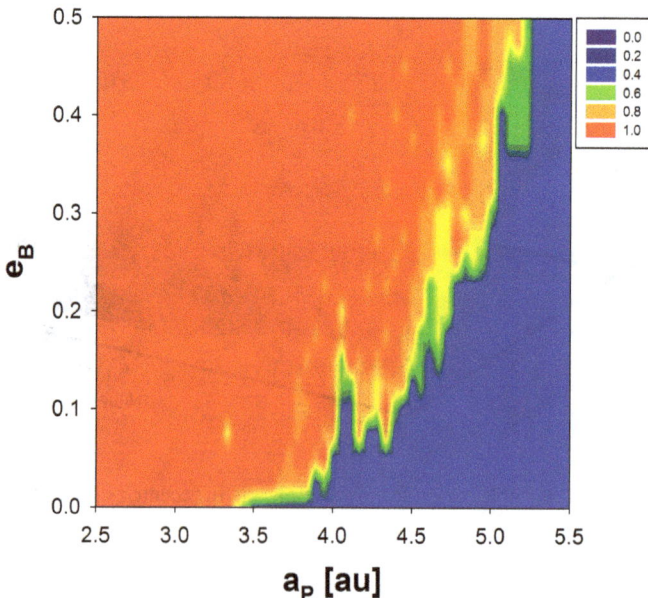

Figure 2.9: Stability of eccentric planetary P-type motion around two G-type stars for different eccentricities of the binary. The initial eccentricity of the planet is 0.3 and the colors represent its maximum eccentricity in the simulation. For details see the text.

remaining stable area is only 20% of that for circular planetary motion, as defined by general stability studies using the RTBP (see e.g., Holman and Wiegert, 1999).

2.3.2 *P-type motion*

In order to highlight the influence of eccentric P-type motion on the stability region shown in Figure 2.5, the same orbits were calculated with an initial eccentricity of $e_P = 0.3$ for the planetary motion. These results are shown in Figure 2.9. A comparison of both studies shows a significant increase of the unstable (red) and the intermediate (yellow and green) areas. For eccentric P-type motion around two G-type stars, where the stellar distance is defined as unity, the stability border of planetary motion has moved outwards from 2.3 au to 3.5 au for circular binary motion. In case of eccentric binary motion ($e_B = 0.5$) the stability border is beyond 5 au. This result indicates a quite strong influence of the planet's eccentricity on the stability border for P-type motion, which can be crucial for the stability of planetary orbits in the HZ.

Chapter 3

Perturbations in Multi-Planet Binary
Star Systems

In this chapter, we review how the mutual gravitational interactions between celestial bodies can lead to the phenomenon of resonances. Resonances are important since they can severely influence the habitability of planets. As part of this review we will recapitulate some established results from the literature. In particular, we focus on methods to determine the location of various kinds of resonances, and especially on a novel method for secular resonances that was introduced recently (Pilat-Lohinger *et al.*, 2016; Bazsó *et al.*, 2017).

Generally, we speak of a resonance when the ratio of two frequencies f_1 and f_2 can be expressed as a rational number, i.e., $|f_1/f_2| = p/q$, where p and q are integers. We can recast this relation into the condition for an exact resonance $qf_1 - pf_2 = 0$, which has important consequences when viewed in the light of the disturbing function and of small divisors.

Depending on the frequencies that are involved, every resonance (or critical angle of a resonance) has a typical time-scale. We can distinguish between *fast* angles (with small periods) and *slow* angles (with correspondingly larger periods). Taking our Solar System as an example, fast angles are related to orbital frequencies, i.e., to the orbital motion of planets and minor bodies around the Sun. These time-scales for the planets range from 0.24 years for Mercury to 165 years for Neptune, but can reach up to thousands of years for objects in the Kuiper belt and beyond. The slow angles are related to precession frequencies, i.e., to variations of the line of apsides and line of nodes. The corresponding *secular* time-scales for the planets range from $\sim 5 \times 10^4$ years for Saturn to $\sim 2 \times 10^6$ years for Neptune (see Laskar, 1988); these periods are typically 10^4 times longer than the orbital period.

We distinguish between different types of resonances depending on the kind of frequencies that are involved. Resonances between two orbital frequencies are called mean motion resonances (MMR; see Gallardo, 2006). There are also cases in which the orbital frequencies of three bodies are resonant, these are named three-body resonances (TBR; see Nesvorný and Morbidelli, 1998). A famous example for this situation are the Galilean moons of Jupiter, where the orbital periods of Io, Europa and Ganymede have the ratio 4:2:1 and form a Laplace resonance. Resonances between two precession frequencies are termed secular resonances (SR). Another type of resonance that can occur is the spin-orbit resonance. In such a case, the rotational (spin) frequency of a celestial body is commensurable with its orbital frequency, meaning that the ratio of the two frequencies can be expressed as the ratio of two integers (Lemaître, 2010). Mercury is a well-known example for it is locked into a 3:2 spin-orbit resonance, meaning that the planet performs three complete rotations for every two orbits around the Sun.

3.1 Mean Motion Resonances

In order to fully understand the role and effect of an MMR, we will introduce the disturbing function to demonstrate the occurrence of resonant angles. For a more in-depth discussion on resonances, and in particular MMR, see Lemaître (2010).

3.1.1 *The disturbing function*

In an MMR, the orbital frequencies of two bodies are involved. This means that we can study MMR in the three-body problem consisting of a central body with mass M and two smaller bodies with masses m and m'. Using the Hamiltonian formalism, we write the system's Hamiltonian function as $\mathcal{H} = \mathcal{H}_0 + \mathcal{R}$. Here, $\mathcal{H}_0$ is the unperturbed (two-body) part of the Hamiltonian, that describes the motion of any of the smaller bodies about the central body in the absence of any other perturbing bodies. The expression $\mathcal{R}$ contains the interaction terms between the two bodies and is called the effective perturbing potential or disturbing function. It is generally written in terms of orbital elements,[1] a, e, i, ϖ, Ω and λ, with a similar set with primed elements for the body with mass m'. The disturbing function $\mathcal{R}$ can be expanded into a Taylor-Fourier series

$$\mathcal{R} = \mathcal{G}m' \sum_{j_n} S(a, a', e, e', i, i') \cos\left[\varphi(\lambda, \lambda', \varpi, \varpi', \Omega, \Omega')\right]. \quad (3.1)$$

The Fourier part contains only cosine terms of the argument φ, where each φ is a linear combinations of the angles

$$\varphi = j_1\lambda + j_2\lambda' + j_3\varpi + j_4\varpi' + j_5\Omega + j_6\Omega',$$

and has to satisfy the d'Alembert rules (see Morbidelli, 2002), which require that the sum of all coefficients j_n equals zero. Note that the sum in Equation (3.1) is over all six integer indices j_n ($n = 1 \ldots 6$). The Fourier coefficient S of each cosine term is a power series of the indicated elements. The disturbing function $\mathcal{R}'$ for the body with mass m' has in principle the same form with respect to the orbital elements, just with $\mathcal{G}m'$ replaced by $\mathcal{G}m$.

There are different ways to explicitly express the disturbing function in Equation (3.1). Some variants use a power series expansion for small values of e and i and keep the exact dependence on $\alpha = a/a'$ in the form of Laplace coefficients (Ellis and Murray, 2000), while others use an expansion in powers of α, but keep the exact functional dependence on e and i (Laskar and Boué, 2010; Mardling, 2013). Still other forms exist that do not work with the usual heliocentric coordinates, but express $\mathcal{R}$ in other coordinates (Laskar and Robutel, 1995).

3.1.2 *The small divisor*

For a deeper insight into the theoretical meaning of MMR, we have to make use of Equation (3.1). When there is an MMR, a specific combination of the

---

[1]Here $\varpi = \omega + \Omega$ denotes the longitude of pericenter, and $\lambda = \omega + \Omega + M$ is the mean longitude. The other elements are the same as described in Chapter 2.

mean longitudes $(j_1\lambda + j_2\lambda')$ will be close to zero. We isolate the term $\mathcal{R}_{j_1,j_2}$ from the disturbing function and write it as $\mathcal{R}_{j_1,j_2} \propto \cos\left[j_1\lambda + j_2\lambda' + \beta\right]$, where $\beta = j_3\varpi + j_4\varpi' + j_5\Omega + j_6\Omega'$ is a phase angle that depends on the indicated indices and variables. After substituting the expressions $\lambda = nt + \varpi$ and $\lambda' = n't + \varpi'$ into $\mathcal{R}_{j_1,j_2}$ from above, we see the explicit time dependence

$$\mathcal{R}_{j_1,j_2} = \mathcal{G}m'S\cos\left[(j_1n + j_2n')t + \tilde{\beta}\right], \tag{3.2}$$

where $\tilde{\beta}$ incorporates the extra terms in ϖ and ϖ'.

The important point here is that we need to insert this part of the disturbing function into Lagrange's planetary equations to obtain the time-variation of the orbital elements. Once we have established the differential equations for the time-variations, the next step is to integrate them to obtain the orbital elements as functions of time. As an example we show how Equation (3.2) affects the eccentricity, for which the corresponding equation is $de/dt \propto \partial \mathcal{R}/\partial \varpi$. Evaluating the partial derivative of Equation (3.2) does not cause any problem, but to solve for $e(t)$ we have to perform an integration with respect to time. After performing this integration we obtain

$$e(t) \propto \mathcal{G}m'S\frac{\partial \tilde{\beta}}{\partial \varpi}\frac{\cos\left[(j_1n + j_2n')t + \beta\right]}{j_1n + j_2n'}.$$

In this way we clearly see that the small divisor $j_1n + j_2n' \approx 0$ leads to large variations of the eccentricity (in this example) whenever the frequencies n and n' form an integer ratio $n'/n = |j_1/j_2|$. Therefore it is important to know the locations of MMRs in a planetary system.

3.1.3 *Resonance location*

Let us assume that two bodies with masses m and m' are in an MMR and that we know the position of the body with mass m'. We can then derive the necessary condition for the resonant semi-major axis, a_{res}, of the other body from Kepler's third law. This condition can be stated in the following way:

$$a_{\text{res}} = a'\left(\frac{n'}{n}\right)^{2/3}\left(\frac{M+m}{M+m'}\right)^{1/3}. \tag{3.3}$$

We have to distinguish two different cases:

(1) If $n'/n < 1$, then it follows that $a_{res} < a'$ and this is an internal resonance. The body with mass m orbits closer to the central body than the body with mass m'.
(2) If $n'/n > 1$, then it follows that $a_{res} > a'$ and it is an external resonance. In this case, the body with mass m moves outside the orbit of the other body.

In order to quickly visualize these cases one can write $n'/n = p/(p+q)$ for internal resonances, and $n'/n = (p+q)/p$ for external ones. Here, p and q are integers, and q is called the order of the resonance.

3.2 Secular Resonances

As mentioned above, secular perturbations act on time-scales much longer than the orbital (revolution) time-scale, i.e., $T_{sec} \gg T_{rev}$. We follow a similar procedure as before to introduce SR via the averaged disturbing function, and show that small divisors appear when there is a resonance.

3.2.1 *The averaged disturbing function*

The averaging principle states that the short period terms in Equation (3.1) contribute with zero time-average (over the time interval T_{sec}) to the long-term evolution of a dynamical system. This means that we can get rid of all short period contributions by eliminating all terms that depend on the fast variables λ and λ'. In this way, we obtain the averaged (secular) disturbing function

$$\mathcal{R} \longmapsto \langle \mathcal{R} \rangle = \frac{1}{(2\pi)^2} \int\!\!\int_0^{2\pi} \mathcal{R} \, d\lambda \, d\lambda'.$$

By eliminating the mean longitudes, the semi-major axes become time independent and can be treated as constants. In effect, $\langle \mathcal{R} \rangle$ still contains the elements a and a', but only in the form of the semi-major axis ratio $\alpha < 1$.

The remaining variables can be substituted by the so-called Laplace–Lagrange variables (h, k, p, q):

$$h = e \sin \varpi \quad p = \sin(i/2) \sin \Omega$$

$$k = e \cos \varpi \quad q = \sin(i/2) \cos \Omega.$$

In this set of variables, the differential equations describing the time-evolution of the eccentricity and inclination decouple from each other in the lowest order expansion and hence can be solved independently.

3.2.2 Secular evolution of a test particle

Consider a test particle with a negligible mass relative to the mass of any perturbing body. We will present the solutions to its equations of motion following Murray and Dermott (1999). This particle has orbital elements a, e, i, ϖ, Ω and λ, while the other N perturbing bodies have masses m_j and semi-major axes a_j. In the simplest case, the averaged perturbing function is given by

$$\langle \mathcal{R} \rangle = na^2 \left[\frac{1}{2} ge^2 + \sum_{j=1}^{N} A_j ee_j \cos(\varpi - \varpi_j) \right].$$

Note that we neglect the terms related to the inclination (assuming a co-planar system), hence no Ω appears in this function. Here, e_j (ϖ_j) are the perturber's eccentricities (longitudes of pericenter), n is the particle's mean motion, and the important variable g is the particle's proper secular frequency calculated from

$$g = \frac{n}{4} \sum_{j=1}^{N} \frac{m_j}{M} \alpha_j \bar{\alpha}_j b_{3/2}^{(1)}(\alpha_j). \tag{3.4}$$

The frequency g depends on the mass ratio of the perturber's mass relative to the central body mass m_j/M, and on the semi-major axis ratio $\alpha_j = a/a_j$ (when $a < a_j$) or $\alpha_j = a_j/a$ (when $a_j < a$). This ratio also appears as an argument to the Laplace coefficient $b_{3/2}^{(1)}$, for which the leading term in the lowest order expansion gives $b_n^{(k)}(\alpha_j) \propto \alpha_j^k$.

We omit a few intermediate steps that are necessary to express the equations of motion in the variables h and k. Instead, we write directly the solution of these equations

$$h(t) = e_{\text{free}} \sin(gt + \phi) - \sum_{j=1}^{N} \frac{v_j}{g - g_j} \sin(g_j t + \phi_j)$$

$$k(t) = e_{\text{free}} \cos(gt + \phi) - \sum_{j=1}^{N} \frac{v_j}{g - g_j} \cos(g_j t + \phi_j).$$

(3.5)

The constants e_{free} and ϕ follow from the test particle's initial conditions, with g as in Equation (3.4). Additionally, g_j, ϕ_j and v_j are prescribed by the secular solution of the perturber's motion. The variables g_j play an important role, as they determine the eigenfrequencies of the massive bodies (for $N > 1$). When we abbreviate the sums in these equations by $h_0(t)$ and $k_0(t)$, we can define the forced eccentricity $e_{\text{forced}} = \sqrt{h_0^2 + k_0^2}$. From the Equations (3.5) above, it is again obvious that a small divisor appears whenever $g - g_j \approx 0$; we then speak of a linear secular resonance. In such a case, the test particle's proper secular frequency is close to or equals one of the perturber's precession frequencies. As a consequence, the specific amplitude v_j is strongly enhanced, such that the forced eccentricity may become very large.

3.3 New Semi-Analytical Method

An important application of the theory of secular resonances is for binary star systems that contain multiple planets. We consider a binary star system with the stellar components A and B. Here we denote the mass of the planet hosting star by m_A (i.e., the central body, or "primary" star), which has one (or more) planetary mass object(s) with mass(es) m_j ($j = 1 \ldots N$). We require for the distant stellar companion with mass m_B (i.e., the "secondary" star) that it is far enough to allow stable S-type motion around the primary star (as discussed in Chapter 2). However, we do not restrict by any means the mass ratio of the two stars, i.e., the secondary star can be the more massive one.

One of simplest configurations that we can study is the restricted four-body problem consisting of two stars, a giant planet (GP) and a terrestrial planet (TP). A single planet is less interesting from the point of view of habitability, since it is either located in the habitable zone of the primary or not. The perturbations from the secondary star would lead to quasi-periodic oscillations of the planet's eccentricity and inclination, but on average they would not remove the planet from the HZ, provided that it is not too close to the chaotic region. A second planet already introduces a lot of new dynamical features, such as MMRs and SRs, especially if it is a gas giant.

In the following we treat the terrestrial planet as a massless object, that is initially on a circular and coplanar orbit with the other bodies. We outline the semi-analytical method described by Pilat-Lohinger *et al.* (2016) to determine the locations of linear SR and study the dynamical stability regions. For this method, we split the four-body problem into two coupled three-body problems:

(1) In the *binary star–giant planet* system, we need to determine the planet's secular precession frequency g_{GP}. This is done most accurately by extracting the sought value from a numerical integration by means of frequency analysis. Alternatively there also exist analytical methods to estimate the frequency (e.g., Heppenheimer, 1978; Georgakarakos, 2003).

(2) For the *host star–giant planet–terrestrial planet* system, we apply the Laplace–Lagrange secular perturbation theory (Murray and Dermott, 1999) to find the linear SR of the terrestrial planet with the fixed secular frequency of the giant planet.

3.3.1 *Numerical part of the method*

As a first step, we need to calculate the precession frequency g_{GP} of the giant planet induced by the gravitational interaction with the secondary star. The terrestrial planet does not influence the giant planet, since we have assumed a restricted problem. We then have to insert g_{GP} into Equations (3.5), so the more accurately we know its value the better we can determine the position of the SR where $g - g_{GP} = 0$.

From Newton's principle of action and reaction it follows that also the secondary's pericenter will precess due to the exchange of angular momentum with the giant planet. However, the secondary star's precession frequency g_B will be smaller by about the planet-to-star mass ratio, as we can estimate from Equation (3.4). A typical Jupiter-like planet has a mass of $\sim 10^{-3} M_\odot$, whereas even the lowest mass stars have 80–100 times that mass. Accordingly, for a Sun-like secondary star the precession frequency would be $g_B \sim 10^{-3} g_{GP}$. This value is so small that it is very unlikely that any terrestrial planet can enter into a linear SR with the secondary star's precession frequency.

The numerical part of the semi-analytical method consists of a single numerical integration of the binary star–giant planet system. Among the many popular integration schemes, we chose to use the Lie-series method (Hanslmeier and Dvorak, 1984; Eggl and Dvorak, 2010; Bancelin *et al.*, 2012). Other suitable methods include the Bulirsch–Stoer extrapolation scheme (Bulirsch and Stoer, 1966), the high-order Radau scheme (Everhart, 1974), or the modified Mercury integrator for binary star systems (Chambers, 2010).

The output from the numerical integration (Cartesian coordinates or heliocentric orbital elements) is first converted into the Laplace–Lagrange variables h and k (see Section 3.2.1). Subsequently, we perform a Fourier analysis by either a Discrete Fourier Transform (Reegen, 2007) or a Fast Fourier Transform (Frigo and Johnson, 2005). In this step, we determine the frequencies with largest Fourier amplitudes that dominate the dynamical spectrum of the giant planet. Usually the two most prominent peaks are associated with the precession frequencies of the planet (g_{GP}) and the secondary (g_B). Effectively, g_{GP} is the precession frequency averaged over the integration time T,

$$\langle g_{GP} \rangle = \frac{1}{T} \int_0^T g_{GP}(t) dt.$$

In Figure 3.1 we illustrate the process described above with three examples. The data in the plot are taken from a simulation of the existing binary Gliese 86 (see Table 7.1 for details). The top panel displays the time evolution of the variable $k(t)$ for the giant planet. The three different curves correspond to the following values of the secondary star's eccentricity: $e_B = 0.3, 0.4, 0.5$ in black, dark grey and light grey, respectively. In the bottom panel we plot the Fourier spectra for these cases. The peaks for

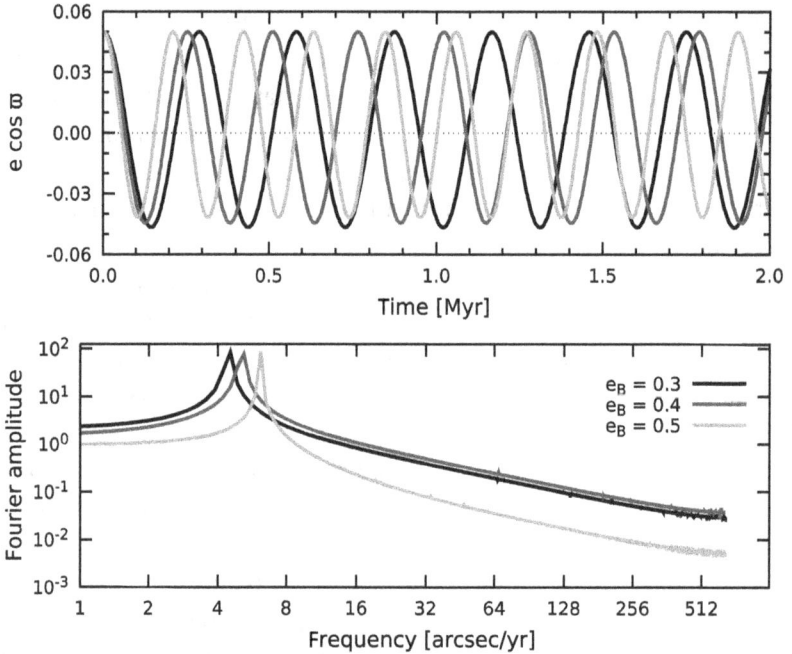

Figure 3.1: Top panel: Examples for the time evolution of $k(t)$. Bottom panel: Their associated Fourier spectra. The three cases differ only by the secondary star's eccentricity e_B.

g_{GP} are clearly visible in all three cases; any peak for g_B would be situated outside the plotted range since $g_B \ll 1$ arcsec yr^{-1}.

It is important to note that the numerical integration has a number of advantages over analytical formulas:

(1) It is not limited by the mass ratio.
(2) The star and planet eccentricities can be large.
(3) The planet frequency can also be calculated for moderate to large semi-major axis ratios α, circumventing the convergence issues in some analytical models.

The disadvantage is clearly that every time the system parameters change we have to perform a new integration of the equations of motion.

Figure 3.2 is directly connected to the second point from above. It is well known that the secondary's eccentricity (e_B) plays an important role and can strongly modify the planet's frequency. In the top panel

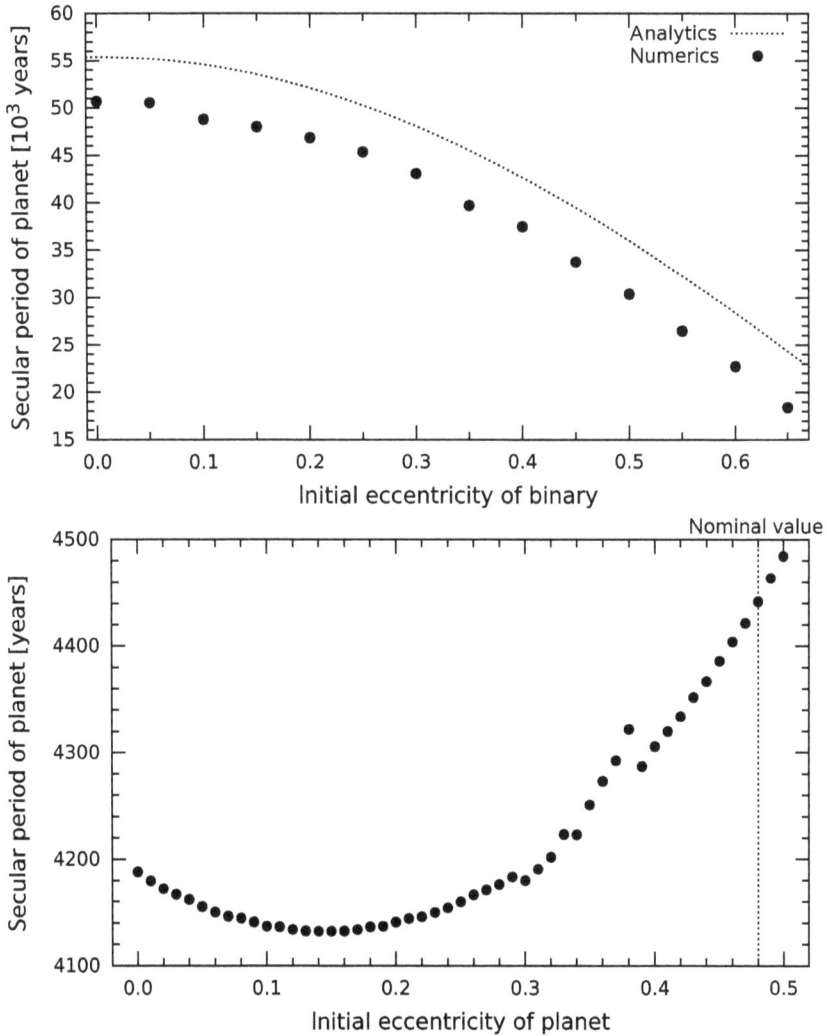

Figure 3.2: Variation of the secular period with the eccentricity of the binary (top) and of the planet (bottom). The black dots are the numerically determined secular periods, while the dotted curve is calculated according to an analytical model (see text for details).

we show the dependence of g_{GP} (in the form of the period) on e_B. The panel shows a synthetic binary of two equal mass G-type stars separated by 60 au with a Jupiter-mass planet located at 3 au. All black points were derived from numerical integrations with a subsequent frequency analysis. The dotted curve represents the analytical model of Heppenheimer (1978). In this pioneering work, Heppenheimer investigated the formation of planets in binary star systems. His goal was to identify regions where planetesimals could successfully grow and accrete to planets under the secondary star's secular perturbations. He employed a restricted three-body problem and derived compact expressions for the secular frequency and forced eccentricity of the massless particle (see Section 6.5 for details about these expressions). As a consequence of neglecting the planet's mass, Heppenheimer's model is known to overestimate the secular period. In Figure 3.2 (top panel) this leads to the vertical offset of the analytical curve (dotted) relative to the numerical data points. In most analytical models, the planet's eccentricity is neglected; however, there is also a variation of a planet's secular period with its initial eccentricity. In the bottom part of Figure 3.2, we show the change in g_{GP} for different initial eccentricities e_{GP} of the planet HD 196885 b (see again Table 7.1). For this planet, the observed value for e_{GP} (nominal value) is rather high ($e_{GP} = 0.48$). The amplitude of the variations is smaller than that caused by varying e_B, but the lowest and highest values are still different by 10%. When the dependence on e_{GP} is not taken into account, we would make an error of about the same magnitude in the location of the SR.

3.3.2 *Analytical part of the method*

For the second part of the method, we use the Laplace–Lagrange secular perturbation theory. It serves as a first approximation for the motion of planetary mass objects with low eccentricity and inclination. These restrictions originate from the limitations of the theory, which is first-order in the masses and only includes terms of degree e^2 and $(\sin i)^2$, while higher order term were neglected.

One can use an iterative scheme that computes the terrestrial planet's precession frequency g according to Equation (3.4) over a suitable semi-major axis interval, and insert that value into Equations (3.5) until the condition for the SR ($g - g_j = 0$) has been met. All necessary parameters (masses, semi-major axes and frequencies of the perturbers) are fixed at this point; the only variable is the position, a_{TP}, of the terrestrial planet.

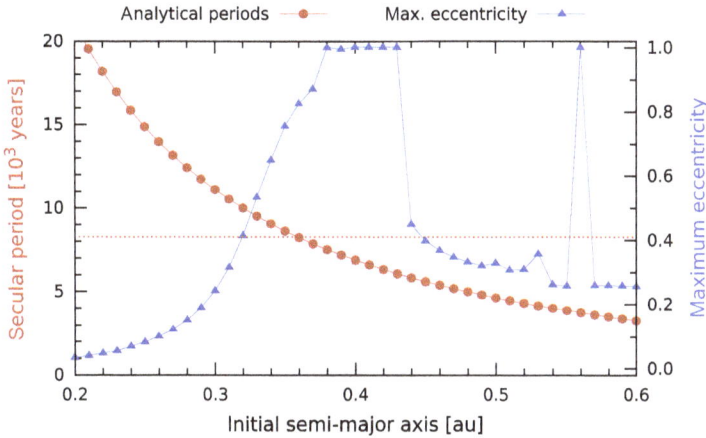

Figure 3.3: Analytical periods (red curve) calculated with the Laplace–Lagrange theory for test planets, and the corresponding maximum eccentricities (blue curve) from a numerical integration. The dotted horizontal line indicates the secular period of the giant planet ($\approx$ 8000 yrs).

Figure 3.3 provides a visualization of this scheme. For the binary HD 41004, we plot the analytical periods of test planets on a regular grid for the interval $0.2 \leq a_{TP} \leq 0.6$ au (red curve with bullets). When this curve intersects the dotted horizontal line, which is equivalent to the giant planet's secular period (for $e_{GP} = 0.2$), we have found the location of the SR. The blue curve with triangle symbols represent the maximum eccentricity of test particles; these values were extracted directly from numerical integrations for comparison. A grey shaded area indicates the peak values in maximum eccentricity, which is the most chaotic region where all test planets are ejected from the system.

3.3.3 *Benefit of the method*

The presented semi-analytical method is both simple, in the sense that it does not require a large amount of computations, and practical for the purpose of locating the SR of a terrestrial planet with a much more massive giant planet. Consequently, we are able to determine both the locations of MMRs and of SRs for circumstellar planetary motion in binary star systems.

Chapter 4

Terrestrial Planet Formation in Binary Stars

The general view of planet formation breaks the process down into different stages. The most common scenario is the *core-accretion model*, which describes planet formation as follows (see also Figure 4.1):

- The gravitational collapse of a huge gas–dust cloud leads to the formation of a proto-star and a rotating disc of gas and dust. In this disc, coagulation of the dust grains into cm-sized particles takes place.
- Accretion and collisional growth of these particles lead to the growth of km-sized bodies known as *planetesimals*.
- Further growth of these planetesimals results in proto-planetary cores, i.e., Moon to Mars-sized bodies, called *planetary embryos*.
- Finally, collisional growth of embryos and the remaining planetesimals leads to the formation of terrestrial planets.

Understanding the various processes for terrestrial planet formation in binary star systems is of high importance. Gas giant planets have to form[1] within the lifetime of the gas disc. The median disc dispersion time was estimated to be between 2 and 3 Myrs according to Williams and Cieza (2011),

---

[1]When embryos reach a certain mass due to the oligarchic growth (see Kokubo and Ida, 1998) a process of runaway gas accretion may lead to the formation of a gas giant planet (see Stökl *et al.*, 2016).

Figure 4.1: A sketch of the different stages of planet formation from a huge gas cloud to a planetary system.

who determined this from observational surveys (e.g., IRAS and Spitzer satellites) of various star forming regions.

The combined perturbations of the secondary star and an already developed gas planet can become quite significant in the case of rather tight binary systems, such as γ Cephei (Hatzes *et al.*, 2003; Neuhäuser *et al.*, 2007), GJ 86 (Santos *et al.*, 2000; Lagrange *et al.*, 2006) and HD 41004 (Zucker *et al.*, 2004), in which massive exoplanets are known to orbit one of the stars. These discoveries have led to a growing interest in understanding planetary formation processes in general.

4.1   Some Aspects of the Early Phase of Planet Formation

In tight binary star systems, the secondary star heavily truncates and possibly distorts the disc, influencing the formation and evolution of planets throughout the various stages of the planet-forming process. The disc truncation is caused by the companion star through gravitational interaction, as was shown by Artymowicz and Lubow (1994) and Savonije *et al.* (1994), where mainly the outer edge of the disc is influenced (Kley and Nelson, 2010; Müller and Kley, 2012). The inner edge is affected in the case of circumbinary discs (Rafikov, 2013). This truncation shortens the lifetime of the disc and consequently limits the period in which gaseous planets can form. During the accretion of dust particles to km-sized planetesimals, the secondary star can modify the shape and orientation of the particle's orbits, which could stop planetary formation (see e.g., Thébault, 2011 and references therein).

Despite the considerable progress this field has seen over the past decades (Marzari and Scholl, 2000; Barbieri *et al.*, 2002; Lissauer *et al.*, 2004; Thébault *et al.*, 2004, 2006; Quintana *et al.*, 2007; Haghighipour and Raymond, 2007; Guedes *et al.*, 2008; Thébault, 2011), many open questions remain, especially with regard to accretion stages where km-sized objects are expected to grow into planetary embryos within a gaseous disc. The outcome of planetesimal–planetesimal collisions is highly sensitive to the collisional velocities (Benz and Asphaug, 1999; Stewart and Leinhardt, 2009). This could hamper the growth of planetesimals by mutual collisions in binary systems, since increased relative velocities lead to a high probability of disruption instead of successful accretion (Heppenheimer, 1978; Whitmire *et al.*, 1998; Marzari and Scholl, 2000; Thébault *et al.*, 2004, 2006).

Recently, Gyergyovits *et al.* (2014) performed a first parameter study to analyze the interplay between a circumprimary gas disc, gravitationally interacting embryos and a distant secondary star. For this investigation they developed a GPU–CPU 2D hydrodynamics grid code that combines hydrodynamic radiative disc computations with highly accurate N-body simulations. This code was used to probe the differences for planetary formation, when taking into account (i) binary–disc, (ii) binary–protoplanet and (iii) the binary–protoplanet–disc interactions. By simulating a coplanar binary–disc system with a grid centered in the primary star, the interplay and evolution of the three components (stars, disc and protoplanets) was investigated for γ Cephei-like configurations. For γ Cephei, the primary's mass, m_A, and the secondary's mass, m_B, are 1.4 $M_\odot$ and 0.4 $M_\odot$, respectively, with a mutual distance of ~20 au and an eccentricity of 0.4. The gas disc of 0.01 $M_\odot$ around the primary extended initially from 0.5 to 8 au with an initial density profile[2] $\Sigma(r) \propto r^{-1}$, where r denotes the distance from the primary. In addition, 2,048 embryos with masses of 0.016 $M_\oplus$ were distributed randomly around the primary. The arrangement of embryo-sized particles implicitly assumes that the planetesimal accretion phase was successful, which is far from being granted.

The investigation by Gyergyovits *et al.* (2014) is of great importance for tight binary star systems, as it showed for the first time that close encounters between protoplanets and their interaction with frequently occurring spiral density waves in the gas disc induced by the secondary star lead to large-scale momentum and energy exchange. These interactions might increase the semi-major axes and eccentricities of the protoplanets to high values and thus inhibit the agglomeration and merging of planetesimals and protoplanets. Moreover, changes in the mass-weighted disc eccentricity and the longitude of the disc pericenter cause variations in the dynamical evolution of the disc (see Müller and Kley, 2012) and might also affect the movement of protoplanets. This may either render planet formation impossible, or provide an environment where agglomeration is not strongly affected by the disc. The study by Gyergyovits *et al.* (2014) revealed the following information:

(i) The evolution of the disc was not strongly affected by the particles when the total particle-to-gas mass ratio was 10^{-2}. However, a phase drift in the disc's mass-weighted eccentricity and argument of pericenter was observed between solutions where the protoplanets' influence is either incorporated or neglected. Moreover, the initially zero eccentricity of the disc increased to maximum values between

---

[2]This simple density distribution places much material outside the Hill sphere of the primary star where it is strongly affected by the perturbations of the secondary.

0.06 and 0.07 and reached 0.03 or 0.035 by the end of the computations (after 100 binary periods) in all the dynamical models they studied.

(ii) The evolution of interacting particles, perturbed by the stars and the gas disc, was studied for the first time with the combined hydrodynamical–N-body code. The eccentricities of the particles underwent strong variations, especially in the fully interacting model (up to ~0.9 in the first 1,000 years). While in the model without a back-reaction from the disc, the evolution proceeded more slowly, and on average the eccentricity remained below 0.4. Moreover, the N-body relaxation processes occurred faster in the presence of a dynamically evolving and interacting gas disc, with average protoplanetary eccentricities almost twice as high as in non-interacting models. It was also shown that the relaxation time depends on the smoothing parameter, which introduces a numerical viscosity and keeps eccentricities at low values. A quiet (low eccentricity) disc damps the particle eccentricities even when only disc gravity is taken into account.

(iii) The growth of particles yielded different collision probability distributions in the fully interacting model. Disruption dominated in this model in both time intervals that have been considered.[3] Mergers were found to be more probable for the early time interval in the model that considered a quiet disc and the model without a back-reaction from the disc, though in the later time interval disruption dominated also in these dynamical models. Therefore, one can conclude that the growth from embryos to planets within a dynamically evolving gas disc is strongly altered by the dynamical evolution of the disc, which leads to a decreased probability for planet formation at least in the inner parts of the gas disc. This fact can be explained by the action of periodically occurring and inward-moving spiral waves in the disc that are exited by the secondary star together with a periodically varying disc eccentricity which leads to excitations of the embryos and thus higher encounter velocities.

---

[3]Gyergyovits *et al.* (2014) compared two time intervals for merging and disruption.

Obviously, there are many parameters that affect the outcome of the simulations, such as the gravitational smoothing parameter (which influences the disc–protoplanet and protoplanet–protoplanet interactions), the type of boundaries (reflecting, outflow, non-reflecting) of the grid in the hydrodynamic part, different flux-limiter functions in the advection part of the code, and the orbit evolution of the secondary star. Therefore, future studies are needed to investigate the influence of the various parameters and shed more light on the problems of the early phase of planet formation in binary star systems.

However, accepting the fact that giant planet formation is possible even in quite tight binaries in the framework of the core-accretion scenario, we may assume that isolated planetary embryos can form as the result of an oligarchic growth of planetesimals. Consequently, a system of terrestrial planets may also form in binary star systems during the final assembly phase of planetesimals. It is a well established fact that giant planets play a decisive role for terrestrial planet formation (Levison and Agnor, 2003). The number of giant planets and their orbital architecture can strongly influence the region where possible habitable terrestrial planets are believed to form (Pilat-Lohinger *et al.*, 2008a,b).

4.2   Terrestrial Planet Formation in the HZ

While gas giant planets have to form quite rapidly within the lifetime of the gas disc, terrestrial planet formation may take some tens to hundreds of millions of years. Collisions of embryos and planetesimals with protoplanets are believed to be the foundation of planetary growth. In binary star systems, the gravitational perturbations from the secondary star can excite the orbital eccentricities of the planetesimals which will increase their impact speeds so that disruption could dominate instead of successfully merging (Heppenheimer, 1978; Whitmire *et al.*, 1998; Marzari and Scholl, 2000; Thébault *et al.*, 2004, 2006). However, observations of giant planets in close binaries, such as γ Cephei, GJ 86 and HD 41004 AB, suggest that terrestrial planets may also form in stellar systems during the final assembly phase of planetesimals.

In this context, numerical simulations by Haghighipour and Raymond (2007) demonstrated that terrestrial planets can be easily formed even in

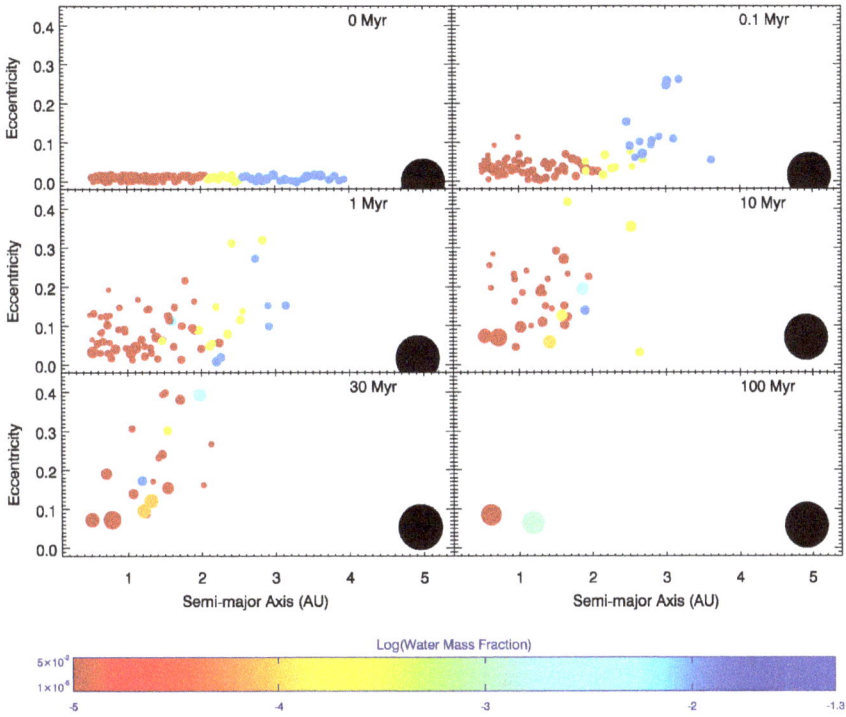

Figure 4.2: Snapshots of terrestrial planet formation in a tight binary where the secondary star (of $0.5 M_\odot$) is at 30 au in an eccentric orbit with $e_B = 0.2$. The evolution of the protoplanetary disc is shown for certain times which display the gravitational interaction in the system until two terrestrial planets have formed after 100 Myrs. The black circle indicates a Jupiter-sized planet. (This figure is taken from Haghighipour and Raymond (2007).)

tight binary star systems. An example of this investigation is presented in Figure 4.2, which shows snapshots of the terrestrial planet formation scenario at different times. The different panels display the gravitational interactions of embryos (initially located in a disc from 0.5 to 4 au) with the Jupiter-mass planet at 5 au (black circle) and a secondary star of 0.5 $M_\odot$ at 30 au. The two stars move in eccentric orbits with $e_B = 0.2$ around their center of mass. The evolution of this system at different time intervals shows clearly the merging of particles by collisions, where after 100 Myr only two terrestrial planets and the Jupiter remained.

Haghighipour and Raymond (2007) performed such numerical simulations for various binary configurations with separations between 20 and 40 au and compared their results with Solar System computations. From this study, they could conclude that terrestrial planets can easily form in such tight binary star systems where the final architecture of a planetary system depends on the mass-ratio, the eccentricity and the semi-major axis of the binary. A comprehensive survey of the influence of the binary's dynamical and physical parameters on terrestrial planet formation is difficult, though. In the study by Haghighipour and Raymond (2007), a big diversity of planetary systems has been found. As an example, we show in Figure 4.3 a summary of possible binary-star–planet configurations for equal-mass stars. In all of these simulations, a prior formed giant planet influenced the

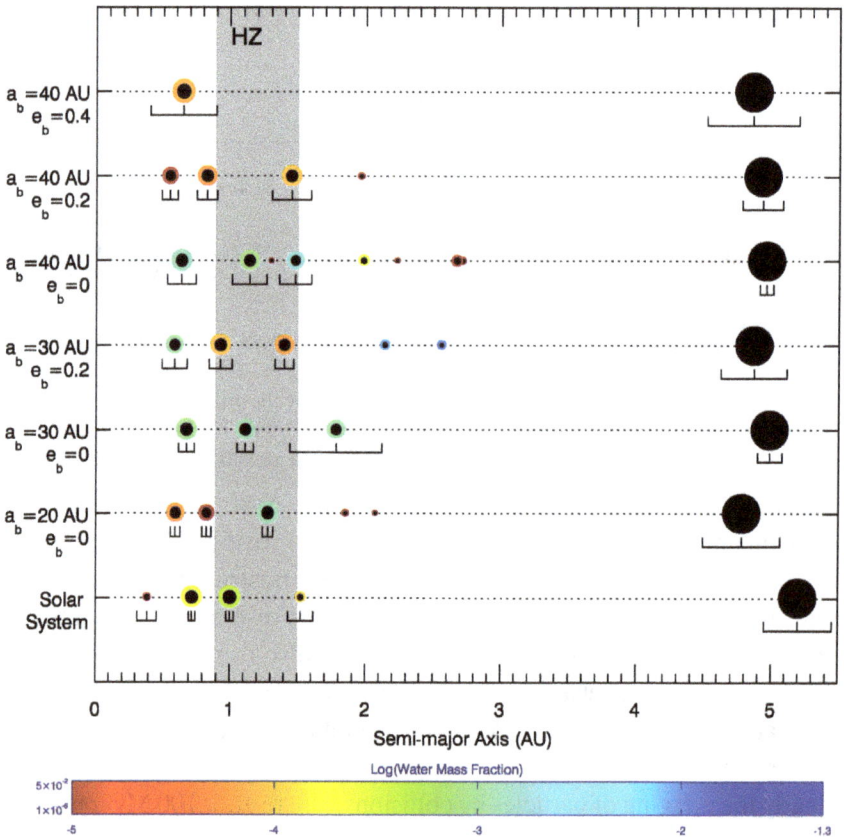

Figure 4.3: Planetary systems formed due to core accretion in tight equal-mass binary star systems. (This figure is taken from Haghighipour and Raymond (2007).)

formation of terrestrial planets significantly, since the giant planet can be considered as the main driver for radial mixing of the planetesimals and embryos in discs.

To emphasize the role of the secondary star and a giant planet for terrestrial planet formation in the HZ more specifically, further numerical simulations of various binary–planet configurations were performed taking into account results from Solar System studies. Therefore, as host star a G2V star with one solar mass ($M_\odot$) was considered which allows to adopt the HZ borders from the Solar System, i.e., between 0.95 and 1.70 au according to the Kopparapu *et al.* (2014) model. This corresponds to the averaged HZ (AHZ, see Chapter 6) in binary star systems. A variation of the mass of the secondary star between 0.4 and 1.5 $M_\odot$ and its distance to the host-star between 25 and 100 au allows one to study the influence of the secondary. In addition, the binary's eccentricity e_B was varied between 0 and 0.6, i.e., from circular to a quite eccentric motion. The numerical simulations were performed with and without giant planets, which revealed the combined influence of the giant planet and the secondary star. For a G2V host-star, the Jupiter-mass planet was typically placed at $a_{GP} \sim 5$ au on a (nearly) circular orbit for the sake of comparison with studies of the Solar System. Of course, a_{GP} was adjusted to closer distances from the host star according to the requirement of dynamical stability of the binary configuration. Studies of pure gravitational interactions of a disc of protoplanets in the various binary star–planet configurations usually show perturbations due to (i) the secondary star, (ii) the giant planet and (iii) the combined effect of both.

In these numerical simulations, the protoplanets are distributed radially with mutual separations depending on their "isolation mass", i.e., embryos have accreted smaller objects in their respective "feeding zone", and each of them has emptied an annular region of several times its Hill radius. The total mass of the embryo population depends on the prescribed surface density profile of solids ($\propto r^{-1}$) which is chosen according to observations of protoplanetary discs. The inner border is set to 0.3 au, while the outer border depends on the semi-major axis in (a_{GP}) of the giant planet. Typically the outer border is 2.6 au for $a_{GP} = 5$ au or 1.6 au when $a_{GP} = 2.5$ au. Consequently, in such a system, all embryos start inside the snow line at 2.7 au (Lecar *et al.*, 2006), which means that the final terrestrial planets necessarily will form dry. If the so formed planet ends up in the HZ, the water has to be delivered to this planet at a later time in order to fulfill basic conditions for habitability.

4.2.1   *Evolution of the protoplanetary disc*

Numerical simulations of protoplanetary discs require an appropriate N-body code[4] that studies the motion of bodies with respect to the system's barycenter. For terrestrial planet formation, one usually traces the orbital evolution of numerous particles with masses of 10^{-8}–10^{-6} $M_\odot$ over a timespan up to 10^8 years.

When collisional fragmentation is excluded, two embryos merge completely once their mutual distance becomes smaller than a pre-defined collision threshold. This assumption is a weak point of such studies and should be improved in future investigations. A more sophisticated model is that of Leinhardt and Stewart (2012) and Stewart and Leinhardt (2012), where different collision regimes — such as hit-and-run, disruption and super-catastrophic disruption — are considered in addition to the perfect merging of bodies. Moreover, for a more realistic formation scenario, it would be necessary to combine the N-body calculations with simulations of collisions, using for example, an SPH[5] code that calculates for different impact velocities and angles how much of the original material in the colliding bodies remains on the final body at the end of the merger.

Typically one or several planets will be formed depending on the parameters of the binary–planet configuration. However, the outcome of individual simulations is strongly stochastic due to a violently chaotic phase in the first few millions of years, as shown in Figure 4.4. In this phase, the orbits of the embryos start to cross as the eccentricities get excited.

---

[4] Such codes include the *nine package* of Eggl and Dvorak (2010), or the *Mercury-Binary* of Chambers (2010).

[5] SPH means Smoothed Particle Hydrodynamics and is a numerical method for hydrodynamical simulations which can also be used for studies of the deformation of bodies — for details see Schäfer (2005), Maindl *et al.* (2013) and Schäfer *et al.* (2016).

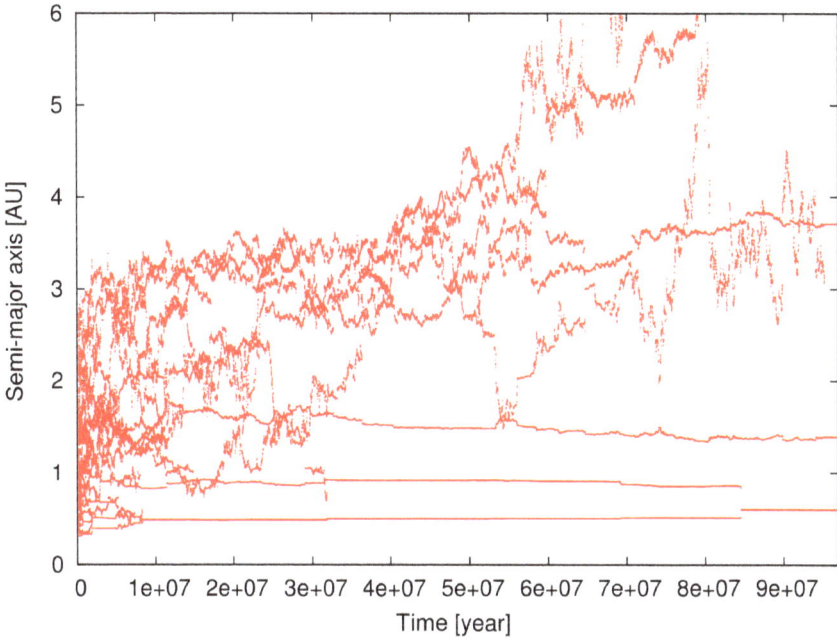

Figure 4.4: Time evolution of embryo semi-major axes in a binary system with parameters $a_B = 100$ au and $e_B = 0.01$ in the absence of a giant planet. By mutual interactions the initial population is expanding on average to larger semi-major axes.

Figure 4.4 shows an example for the time evolution of the semi-major axes of an ensemble of initially isolated embryos in absence of a giant planet. Although embryos start in the region between 0.3 and 2.6 au, they are quickly dispersed outwards to larger distances. This self-stirring of the embryo population by mutual gravitational perturbations and close approaches causes their eccentricities to increase and their orbits to cross. It is also visible from the figure that the growth process is still going on at the end of the simulation since collisions may still happen at later times, see the collision at 85 Myr of two well separated embryos that had both seemed to be on quite stable orbits for a very long time. The influence of the companion star on the embryo dynamics is rather small for moderate and large stellar separations. There would be a significant influence only for large masses or eccentricities of the secondary.

When a giant planet is included in the simulations, we see a very different picture. Figure 4.5 shows two example cases for a 0.4 $M_\odot$

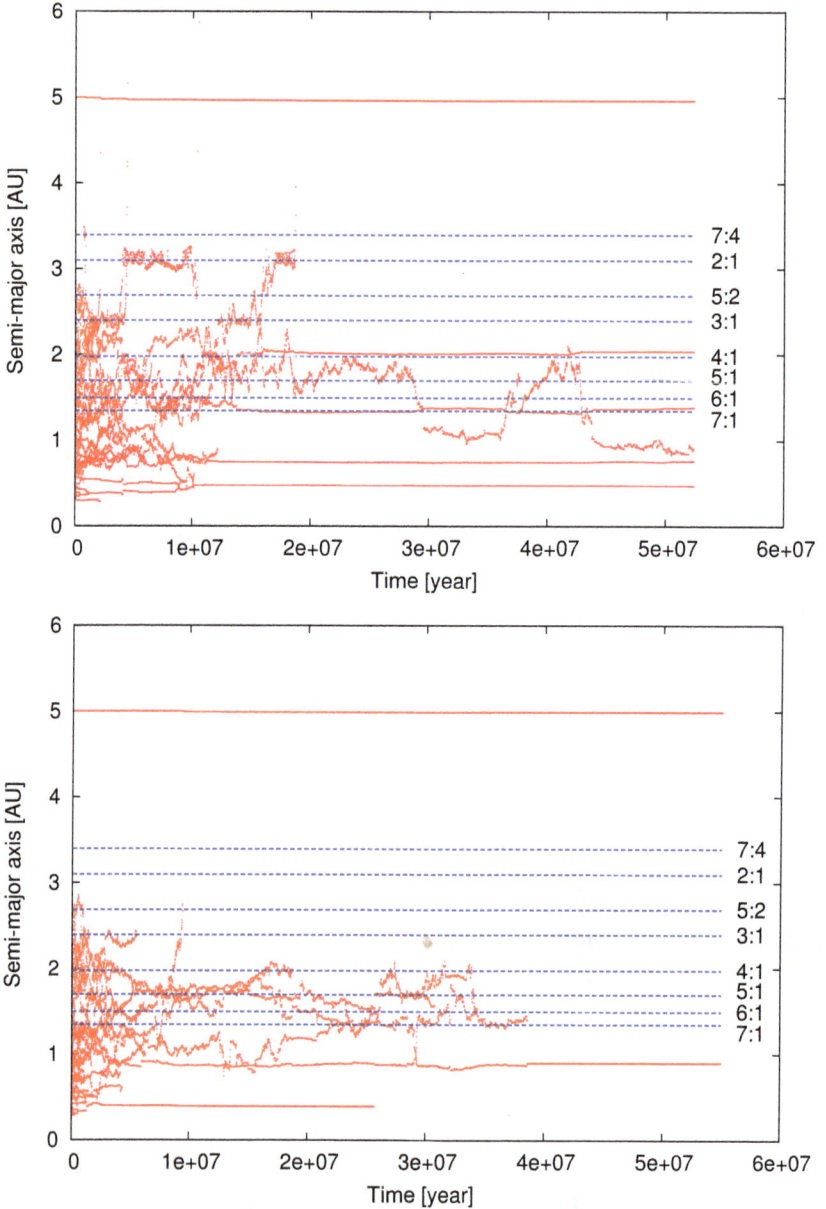

Figure 4.5:   Time evolution of embryo semi-major axes in a binary system with parameters $a_B = 100$ au and $e_B = 0.01$ (top), and $a_B = 100$ au and $e_B = 0.6$ (bottom). The blue-dashed horizontal lines indicate MMRs with the giant planet at 5 au.

companion at 100 au with eccentricity of $e_B = 0.01$ (top) and $e_B = 0.6$ (bottom). The first major difference is that the embryos cannot spread outwards beyond the 2:1 MMR with the giant planet (at about 3 au). Embryos with larger semi-major axes are efficiently scattered out from the system by close encounters with the giant planet. Many objects are affected by MMRs and perform a random walk between several resonances. For a low binary eccentricity (top panel), the scattering occurs at low order MMRs (e.g., 2:1, 3:1 MMR), while for higher eccentricity, also higher order MMRs are affected (e.g., 4:1, 7:1 MMR, bottom panel). Note that for the former case (top panel), there are two objects that stay close to the 4:1 and 7:1 MMRs for considerable periods of time. This result demonstrates that terrestrial planets can indeed form in a binary star system with a giant planet, where in addition to MMRs, SRs may also play an important role. In the lower panel of Figure 4.5, the innermost object escapes suddenly after more than 28 Myrs. This is caused by an SR where the precession frequency of the perihelion is the same as that of the giant planet (see Chapter 3).

4.2.2 *Terrestrial planets in the HZ*

Figure 4.6 shows a comparison of numerous simulations for a binary star system with stars of spectral types G and K, with and without a giant planet of one Jupiter-mass (marked by the smaller red circles in the lower panel). The horizontal axis symbolizes the distance to the host star, with the HZ being the light-blue area located between 0.95 and 1.67 au (like in our Solar System). The giant planet is located either at 2.5 au (for tight binaries), or at about 5 au (like in the Solar System). Both figures show the formation of terrestrial planets in the HZ for all binary separations a_B. The point size indicates the different masses of these planets, which depend on the frequency of collisions and, therefore, on the dynamical perturbations during the formation phase. Comparing the two figures, one can see that in the binary system without Jupiter (upper panel) there is a higher diversity of planetary systems formed after 100 Myr, where even highly eccentric planetary motion (see the horizontal bar) is possible in the outer region. In the lower panel, the giant planet limits the occurrence of terrestrial planets to the region within the 3:1 MMR at about ~2.5 au. Another striking difference is that the orbits of formed planets are more

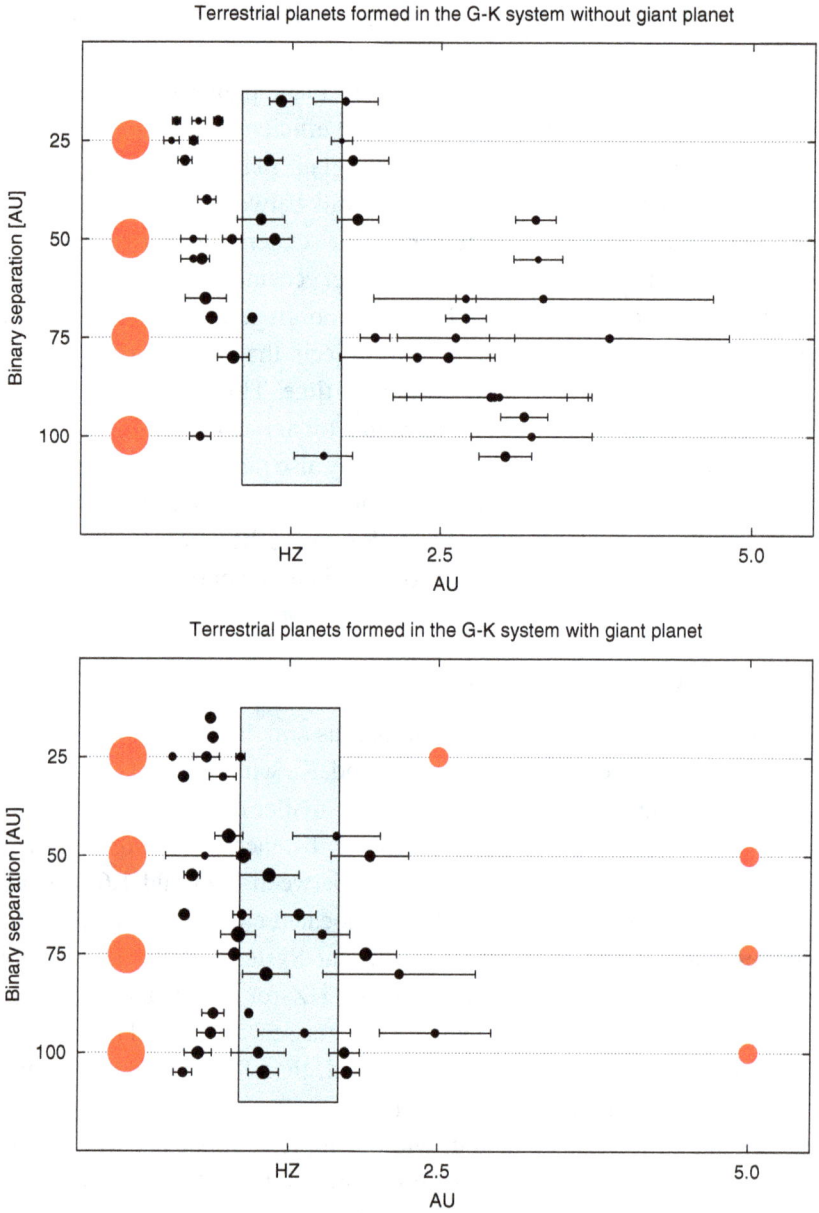

Figure 4.6: Terrestrial planets formed in the HZ of binary star systems when no giant planet is present (top), and with a giant planet marked by the smaller red dot (bottom).

ordered in the latter case. Both panels show the formation of planets in the HZ for which the habitability cannot be guaranteed solely from their locations in the HZ. As we know that a basic requirement for the Earth's habitability is the existence of liquid water on the planetary surface, we need information how the water is transported to a planet in the HZ during the formation phase.

4.3 Water Transport to the HZ

The water transport through icy planetesimals towards the circumstellar HZ of a G2V type star in various binary star configurations was studied in detail by Bancelin *et al.* (2015, 2016, 2017). They assumed stellar separations (a_B) between 25 and 100 au and eccentricities (e_B) between 0 and 0.6. In addition, a Jupiter-mass giant planet is located at $a_{GP} = 5.2$ au and moves on an initially circular orbit about the G2V host star. The disc of planetesimals has been modeled as a ring of 10,000 asteroids with masses similar to main-belt objects[6] in the Solar System. This planetesimal disc has a total mass of 0.5 $M_\oplus$ and contains a water amount of about 200 Earth-oceans[7] when the asteroids have an initial water mass fraction (hereafter *wmf*) of 10% (Abe *et al.*, 2000; Morbidelli *et al.*, 2000); only the smallest objects were dry. In the N-body simulations, the asteroids were randomly distributed between the snow line at 2.7 au (Lecar *et al.*, 2006; Martin and Livio, 2012, 2013) and the border of stable motion in the binary configuration (i.e., $a_{\rm crit}$ according to Holman and Wiegert 1999 and Pilat-Lohinger and Dvorak 2002; for details see Chapter 2). Initial eccentricities and inclinations of the planetesimals were randomly chosen $e < 0.01$ and $i < 1°$. The dynamical evolution of the planetesimals was studied over 10^7 years and resulted in a classification of four groups (Bancelin *et al.*, 2015):

HZ crossers (HZc): Asteroids crossing the HZ
Ejected: Asteroids leaving the system (i.e., semi-major axis $\geq$500 au)

---

[6]Individual objects have masses between 0.001 and 1 Ceres mass, which were determined through 3D-SPH simulations of collisions of basaltic-rocky objects (for details see Bancelin *et al.*, 2015).
[7]An "Earth-ocean" is equivalent to 1.5×10^{21} kg of H_2O.

Collision: Asteroids colliding with the gas giant or the stars
Alive: Asteroids still moving in the belt after 10^7 year.

4.3.1 *Statistical overview*

The statistical overview in Figure 4.7 uses the above classification and summarizes the study of four different binary configurations as examples. In this figure, each bar graph illustrates the fraction of planetesimals that belongs to one of the four groups defined above. Comparing the two panels, for a binary separation of 50 au, a significantly higher percentage of planetesimals cross the HZ (i.e., black histograms on the top) due to stronger perturbations in the planetesimal belt. Moreover, the number of HZc increases significantly when the binary's eccentricity is raised from 0.1 to 0.3. The bar graph for $e_B = 0.3$ in the left panel shows that only 10 to 15% of the planetesimals were still in the belt, about 40% of the asteroids were ejected, and more than 40% collided either with the gas giant or with the host-star. The number of HZ crossers is more than 40% of the initial

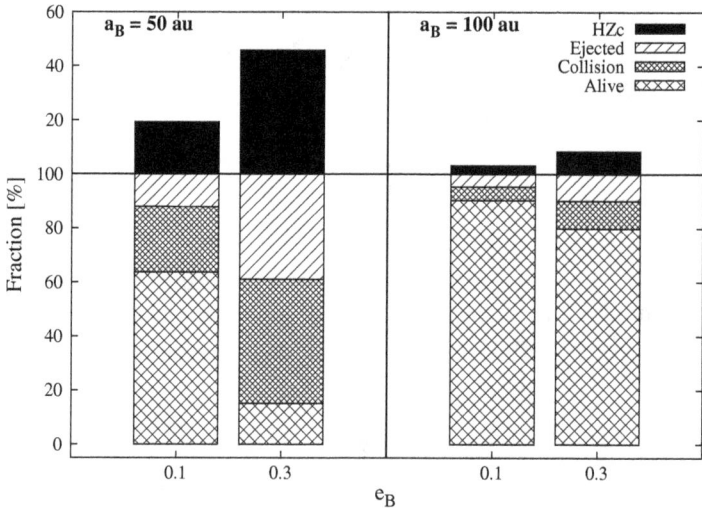

Figure 4.7: Statistical overview of the evolution of the planetesimal disc after 10^7 years for two different binary separations and eccentricities. Each histogram shows the fraction of asteroids that (i) are still in the disc (labeled as "alive"), (ii) have crossed the HZ (i.e., HZc), (iii) have been ejected from the system, and (iv) have collided either with the star or the giant planet.

population; thus there are more than twice as many HZc as in the binary system with lower eccentricity ($e_B = 0.1$). In contrast, the results of the binary systems with a separation of the two stars of 100 au show that most of the planetesimals remained in the belt for the whole integration time and only a few percent collided with one of the massive bodies or were ejected from the system. In addition, the number of asteroids crossing the HZ (see the black bar at the top of an histogram) is quite small. Therefore, the water transport in such systems would not be very efficient within the time of 10^7 years.

This statistical overview indicates that perturbations during the early phases of a planetary system are important for the water transport into the HZ and, therefore, for habitability. Bancelin *et al.* (2016) discussed in detail the stronger perturbations for binaries with $a_B = 50$ au in comparison to a wider binary star system with $a_B = 100$ au where perturbations are essentially near resonances (MMRs and SRs, see Chapter 3 for a description). It is well known from Solar System studies that orbital resonances play an important role for the dynamical behavior in a planetesimal disc (like the Kirkwood gaps in the main asteroid belt). To estimate the gravitational perturbations in a planetary system, it is advisable to calculate the locations of resonances (see Chapter 3) and to study the "dynamical lifetime"[8] of test-planets in these areas as they can be scattered towards the HZ and contribute to the water transport to a planet moving in this area. Bancelin *et al.* (2016) showed that stronger perturbations in a disc lead to a higher flux of asteroids crossing the HZ, where perturbations at the MMRs are triggered by a giant planet which is influenced gravitationally by the secondary star. An increase in eccentricity is connected to an increase in the orbital velocity of the celestial body, which might have consequences for the collisions during the formation and, therefore, also for the water transport.

A further consequence of stronger perturbations in a planetesimal belt is that in tight binary star systems with high eccentricities, the disc will be quickly depopulated. Therefore, we can assume that in such systems the water transport happens more violently and on shorter time-scales than

---

[8]The dynamical lifetime, D_L, is defined as the time when 50% of the test-planets have escaped from the MMR (Gladman *et al.*, 1997).

in wider binary star systems. In the study by Bancelin *et al.* (2015), it was found that a higher binary eccentricity decreases the HZ crossing time significantly.

4.3.1.1   *The role of the secondary's mass*

Results of non-equal mass binary stars (G-K and G-M pairs, where K and M denote the spectral type of the secondary star with 0.7 and 0.4 $M_\odot$, respectively) are displayed in Figure 4.8. The upper panel shows the statistical overview for a stellar separation of 50 au, with the lower panel showing the same overview for a wider binary with the secondary at 100 au. A comparison of the various configurations in both figures indicates clearly that a decrease of the secondary's mass to a K or an M star decreases the number of HZ crossers (black box on the top of each bar graph) and more planetesimals remain in the disc. Therefore, one can conclude that the higher the mass of the perturbing star and the smaller the minimum distance of the two stars (periapse), the stronger are the perturbations of the planetesimals in the disc, which increases the flux of planetesimals towards the HZ and thus increases the water transport.

4.3.2   *Water delivery due to collisions*

The probability of collisions between icy asteroids and Moon- to Earth-sized target planets (TPs) in the HZ can be studied using the MOID (Sitarski, 1968). The MOID is the **M**inimum **O**rbital **I**ntersection **D**istance which is determined analytically and is defined as the closest distance between two objects on Keplerian orbits regardless of their real position on their trajectories. To use the MOID, two separate evolutionary N-body simulations are needed: one for the asteroid belt to study the probability of HZ crossers, and the second for the orbital behavior of TPs in the HZ. The results of these two integrations are used in combination to calculate the MOID. A collision occurs if the MOID is comparable to the physical radius of the TP. Using this method, one can estimate the amount of water delivered to a TP in the HZ according to a five-step algorithm described in Bancelin *et al.* (2017).

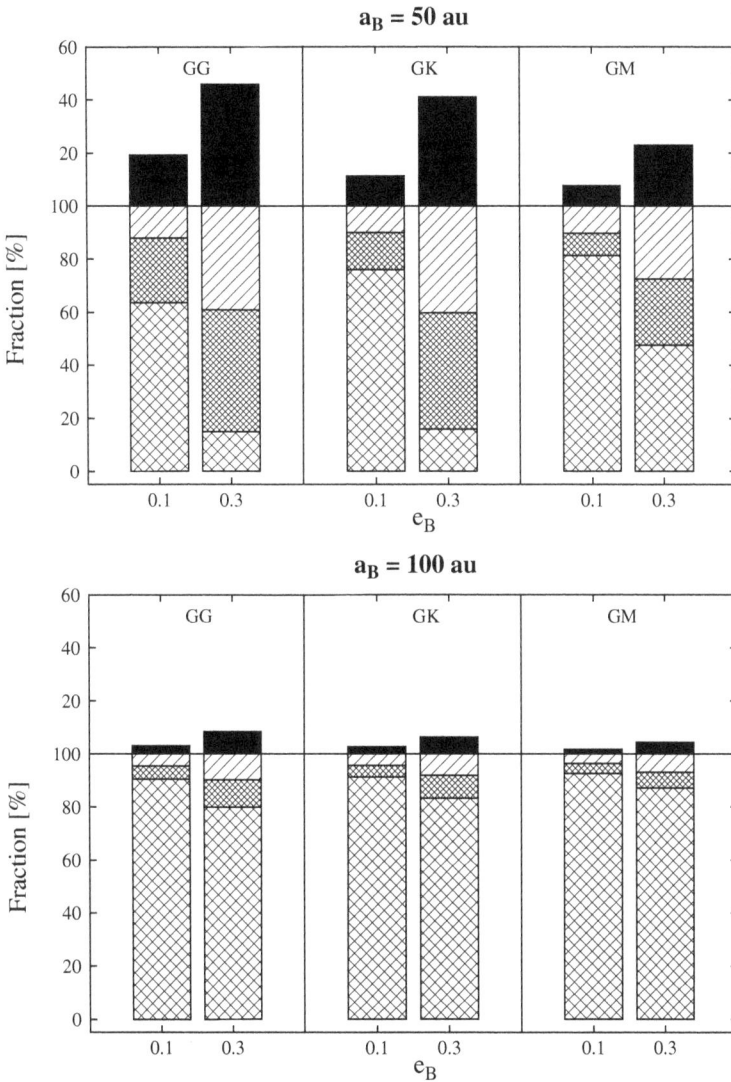

Figure 4.8: Statistical overview of the evolution of the planetesimal disc over 10 Myrs for different binary separations (upper panels: $a_B = 50$ au, lower panels: $a_B = 100$ au) and eccentricities (0.1 and 0.3). Each histogram shows the fraction of asteroids (from bottom to top) that (i) remain in the disc (labeled as "alive"), (ii) have collided with the star or the giant planet, (iii) have been ejected from the system, or (iv) have crossed the HZ (i.e. HZc). The symbols are the same as in Figure 4.7.

In this algorithm, one has first to check if there is an intersection of the TP's orbit with that of an asteroid. Since the TP's eccentricity, e_{TP}, has periodic variations between its initial and maximal value, a function is defined which yields the minimum distance d_{min} between the two orbits, which is the MOID. In case of a collision (i.e., when d_{min} is smaller than the physical radius of the TP), the MOID provides the positions of the TP and asteroid (via their true anomalies) such that their relative impact-velocity and impact-angle can be computed.

The amount of water, $\mathcal{W}_k(a_{TP}, t)$ in units of Earth-oceans ($M_{H_2O} = 1.5 \times 10^{21}$ kg), transported to a TP in the HZ is defined by the asteroid's mass (M_A) and its water mass fraction (wmf_k), where the index k denotes the kth asteroid:

$$\mathcal{W}_k(a_{TP}, t) = \frac{M_A \times wmf_k}{M_{H_2O}}. \tag{4.1}$$

If one sums up $\mathcal{W}_k(a_{TP}, t)$ of all impactors, one gets the total fraction of water delivered to a TP by:

$$\mathcal{W}_{TP}(a_{TP}, t) = \frac{1}{\mathcal{W}_{TOT}} \sum_{k=1}^{N_i} \mathcal{W}_k(a_{TP}, t),$$

where N_i is the number of impactors.

In Table 4.1, we show the water transported to a TP for several binary–planet configurations (see also Bancelin *et al.*, 2017). A comparison of the different systems shows that $\mathcal{W}_{TOT}$ varies significantly in the case of eccentric motion of the TP, while for nearly circular motion the binary's semi-major axis and eccentricity has only a minor influence.

Table 4.1:   Total amount of water $\mathcal{W}_{TOT}$ (expressed in multiples of Earth-oceans) contained in the icy asteroid belt.

a_B	e_B	$\mathcal{W}_{TOT}$ [Earth-oceans] ($e_{TP} \neq 0$)	$\mathcal{W}_{TOT}$ [Earth-oceans] ($e_{TP} \sim 0$)
50	0.3	58.6	78.2
50	0.1	60.2	78.2
100	0.3	75.6	81.5
100	0.1	76.7	81.5

4.3.3 *Water loss during collisions*

Many studies of terrestrial planet formation used perfect merging for the growth of planets, where the entire mass of planetesimals and the entire water content of an asteroid is assimilated into a TP. Such perfect merging scenarios fail to consider any water loss process, such as atmospheric drag or sublimation when the asteroid approaches the host-star. Therefore, it is questionable whether or not $\mathcal{W}_{TP}$ gives a decent estimate. Since we have to assume that some fraction of the water of an icy asteroid will be lost during the collision process, a water loss factor ω_c has to be added to Equation (4.1). This leads to the corrected amount of water, $\tilde{\mathcal{W}}_k(a_{TP}, t)$, given by

$$\tilde{\mathcal{W}}_k(a_{TP}, t) = \frac{M_A \times wmf_k \times [1 - \langle \omega_c(a_{TP}) \rangle]}{M_{H_2O}}, \tag{4.2}$$

where $\langle \omega_c(a_{TP}) \rangle$ denotes the mean water loss factor averaged over all possible impact geometries, $\bar{\theta}_i(a_{TP})$ and impact velocities, $\bar{v}_i(a_{TP})$. The sum over all $\tilde{\mathcal{W}}_k(a_{TP}, t)$ gives a more accurate water fraction delivered to the TP

$$\tilde{\mathcal{W}}_{TP}(a_{TP}, t) = \frac{1}{\mathcal{W}_{TOT}} \sum_{k=1}^{N_i} \tilde{\mathcal{W}}_k(a_{TP}, t).$$

Recently, Bancelin *et al.* (2017) studied the water loss of icy planetesimals during collisions with a TP in the HZ. They used the more realistic collision model and calculated the water loss (ω_c) with detailed simulations of water-rich Ceres-sized asteroids (with a wmf of 15%), colliding with dry TPs with masses of one Moon-, Mars-, and Earth-mass, respectively. For the collision scenario, a 3D SPH code (Maindl *et al.*, 2013) was used, where the key parameters are the impact angles ($\bar{\theta}_i(a_{TP})$) and the impact velocities ($\bar{v}_i(a_{TP})$) of the planetesimals.

4.3.3.1 *Variation in impact velocities and impact angles*

Changes in the impact velocities $\bar{v}_i(a_{TP})$ and impact angles $\bar{\theta}_i(a_{TP})$ — see definitions in Maindl *et al.* (2013) — of planetesimals and TPs due to eccentric motion were investigated in Bancelin *et al.* (2017). An example from that study is shown in Figure 4.9, where the distributions of impact angles $\overline{\theta}_i$ (top panels) and impact velocities $\overline{v}_i$ (bottom panels) are plotted for two binary configurations. The results in Figure 4.9 for the impact angles show strong variations of the two curves near resonances, depending on the

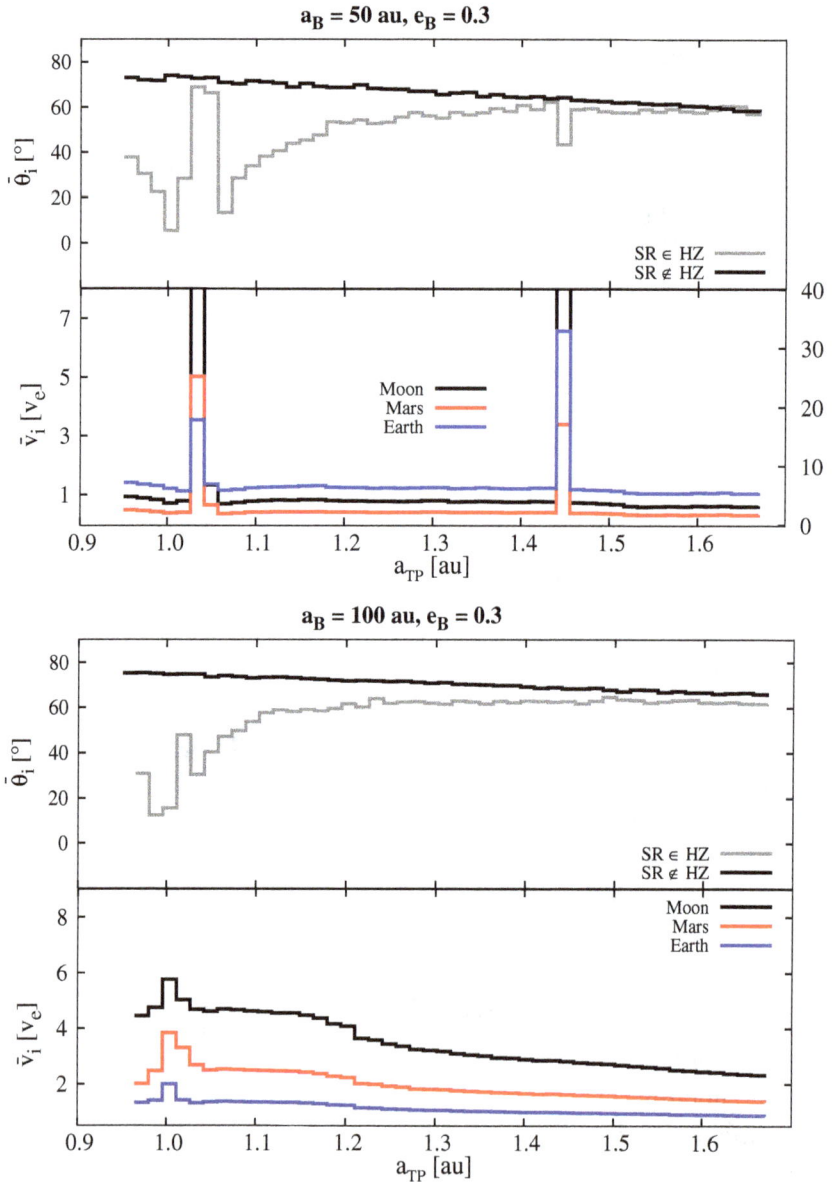

eccentricity of the TP. For nearly circular (i.e., low eccentricity) motion of the TP the black curves indicate impact angles between 60–80°. If an SR or MMR causes eccentric motion of the TP in the HZ, the impact angle $\bar{\theta}_i$ varies between 5–70° (see the light grey curves in the upper panels). Similar differences are also visible for the impact velocities (coloured lines) in the lower panels. In these graphs, impact velocities for collisions on Earth- (blue line), Mars- (red line) and Moon-sized objects (black line) are shown as values normalized to the escape velocity. It is visible that highly eccentric motion near resonances causes significantly higher impact speeds, where $\bar{v}_i$ can reach up to ~$6.0\,v_e$ for an Earth-size body, ~$25\,v_e$ for Mars-size bodies, and beyond $40\,v_e$ for Moon-size bodies.

Detailed SPH simulations are required to understand the impact of the observed range of values in $\bar{v}_i$ and $\bar{\theta}_i$ on the water transport. Figure 4.10 shows some examples of the water loss ω_c during collisions resulting from such SPH simulations. Almost all scenarios result in a merged main survivor retaining most of the mass. We can observe water loss rates between 11% and 68% from this figure.

Using a linear extrapolation of ω_c between the minimum and maximum impact velocities derived for each TP, it is possible to provide better estimates for the water transport. In Figure 4.11 we compare the fraction of water transported to a TP with and without taking into account the detailed collision physics. The total amount of water ($\mathcal{W}_{TP}$) is represented by the dotted lines and the reduced amount of water ($\tilde{\mathcal{W}}_{TP}$) due to water loss is indicated by the solid lines. In addition, different binary–planet configurations are compared, where it is assumed that either orbital resonances perturb the HZ (upper panels) and cause highly eccentric motion of a TP (i.e., around 1 au and between 1.4 and 1.5 au), or the HZ is unperturbed (lower panels). The two top panels show results for

---

Figure 4.9: Impact angles $\bar{\theta}_i$ (top) and impact velocities $\bar{v}_i$ (bottom) with respect to a_{TP} for stellar separations a_B of 50 and 100 au and a fixed $e_B = 0.3$. Colors in the bottom panels are for impact velocities on the surface of Earth (blue line), Mars (red line) and Moon (black line). Two cases are considered, whether an SR lies inside the HZ or not. For $a_B = 50$ au, values of $\bar{v}_i$ for a Moon/Mars TP are on the right side y–axis as these values are much larger than that of Earth (left y–axis). In the top panels, the impact angle for an SR inside the HZ is shown by the grey line and an SR inside the planetesimal belt is represented by the solid black line.

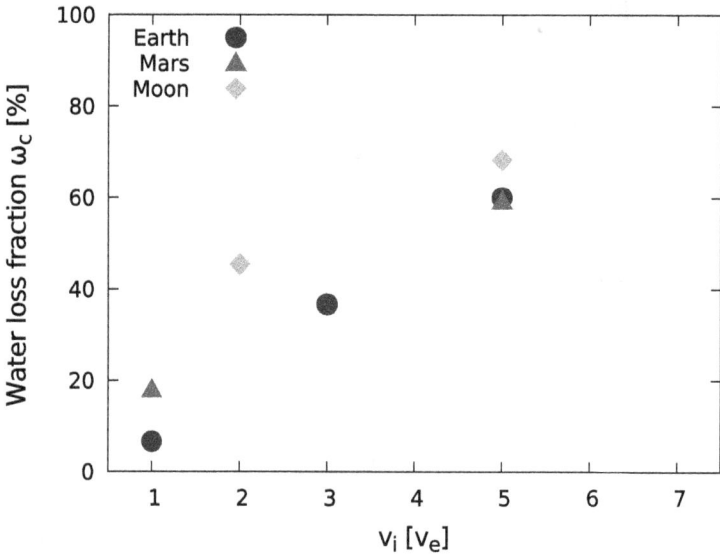

Figure 4.10: Water loss ω_c after impacts as a mass-fraction of the initial total water on Earth-, Mars- or Moon-sized objects for different impact velocities $\bar{v}_i$ in units of the TP's escape velocity v_e.

Moon- and Mars-sized TPs, while the bottom panels are for Earth-sized TPs. For the binary system, the following parameters were used: $a_B = 50$ au, $e_B = 0.3$, and a_{GP} was varied between 3.5 au and 5 au. The evolution of the systems was studied for about 10^8 years. The results show clearly that the water transport by collisions between TPs and asteroids would be severely overestimated if one ignored water loss during this process. This is especially true at positions of SRs (at ~1 au) and MMRs (between 1.4 and 1.5 au) where highly eccentric motion causes a higher frequency of collisions which might pretend a water-rich planet in this area. Taking into account the water loss of icy planetesimals, the transported amount of water to a TP is reduced significantly by almost 50% for an Earth-, 68% for a Mars-, and 75% for a Moon-sized object.

From this study we can conclude that eccentric motion of a TP in the HZ increases the collision frequency, which could increase the water transport to the TP's surface. However, due to higher impact velocities for eccentric motion, the water loss is significantly higher for the icy planetesimals which

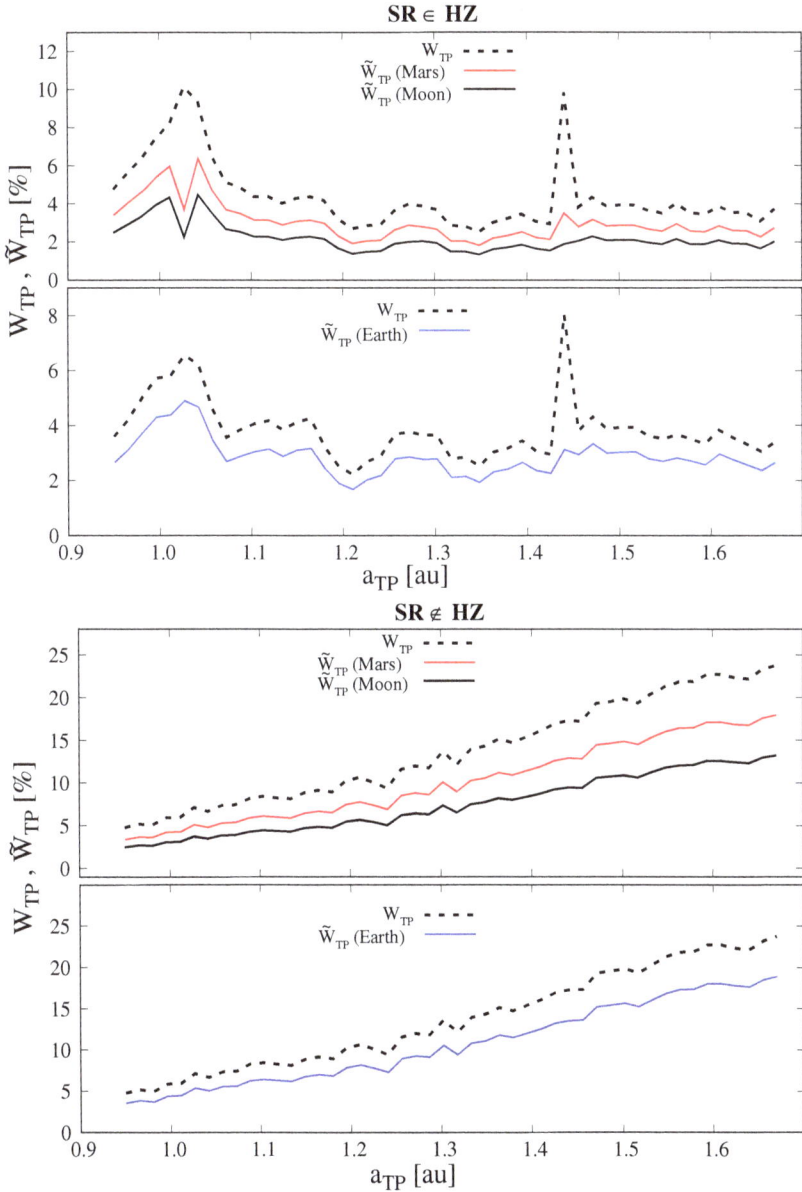

Figure 4.11: Fraction of incoming water with (solid lines) and without (dotted lines) taking into account water loss during collisions. Top panels show the study for a perturbed HZ (by an SR and MMR) where lower eccentricity motion could occur and lower panels show the same for an unperturbed HZ.

drastically reduces the amount of water gained from collisions. Therefore, a result similar to that of nearly circular motion in the HZ can be observed.

It seems that nearly circular motion in the HZ and perturbations (resonances) in the planetesimal belt might be more efficient for the water transport and, therefore, for the habitability of a planet than eccentric motion of a TP in the HZ. Figure 4.11 illustrates this point, as it shows significantly higher values of $\mathcal{W}_{TP}$ and $\tilde{\mathcal{W}}_{TP}$ for an unperturbed HZ (SR $\notin$ HZ, two lower graphs) than for a perturbed one (SR $\in$ HZ, two upper graphs). In any case, SPH collision studies should be included into dynamical studies to avoid misleading effects, as shown by the dotted lines in all panels of Figure 4.11, especially note the spikes at MMRs and SRs in the upper panels. For more details, we refer the reader to Bancelin *et al.* (2017).

4.3.4   *Water transport statistics: Binary versus single star*

The delivery of water to Earth is of high importance when studying habitability on terrestrial planets. Morbidelli *et al.* (2000) investigated the origin of Earth's water and found that our Earth accreted water during its entire formation time. When the formation was completed, there was in addition a late veneer phase when icy planetesimals bombarded Earth. They found that $\sim$10% of the present amount of water was delivered to Earth during this phase.

The previous sections describe water transport in binary star systems after the formation of the planet, which is analogous to water transport during the late veneer on Earth, which could be very important if a terrestrial planet is formed dry.

The role of a secondary star in this scenario was studied by Bancelin *et al.* (2017). They considered a planetary system consisting of two planets moving around a G-type star which is perturbed by a second equal-mass G-type star at 50 au. The binary's eccentricity (e_B) was varied between 0.1 and 0.3. The secondary star perturbs the motion of the Jupiter orbiting the host-star at $\sim$5 au, which therefore changes from an initially circular orbit to an eccentric one. This resulting eccentricity for Jupiter was also used for the computations in the single star system. The size of the disc is also influenced by the secondary star, where the outer border of the disc is defined by the stability limit a_{crit}, that depends on the semi-major axis and eccentricity of the secondary star in the binary system, and the inner border of the disc is the snow-line which is at 2.7 au in the Solar System. Consequently, the discs used for the numerical simulations extend from 2.7 to 12 au for $e_B = 0.1$, and from 2.7 to 9 au for $e_B = 0.3$.

Taking a G-type host star has the advantage that the HZ borders of the Solar System can be applied, which is from 0.95 to 1.67 au according to Kopparapu *et al.* (2014). These values are known as classical HZ and correspond to the AHZ (averaged HZ) defined by Eggl *et al.* (2012), which is described in Chapter 6. In this study, the HZ was divided into four sub-HZs, each having a size of about 0.18 au; they are marked as the inner HZ (IHZ), the inner and outer central HZs (C_1HZ and C_2HZ), and the outer HZ (OHZ). Splitting up the HZ into four sub-HZs gives more details about the locations of impacts when asteroids cross the HZ.

Figure 4.12 shows the total amount of water transported into the HZ (expressed in terrestrial ocean units); the various systems are represented by the different vertical bars. Bars marked with "*B*" at the top refer to binary star systems and those marked with "*S*" show the results for single star systems. The different colors indicate the four sub-HZs where "green" labels the IHZ, "red" labels the C_1HZ, "magenta" labels the C_2HZ and "blue" labels the OHZ. A comparison of a certain binary-star–planet configuration with

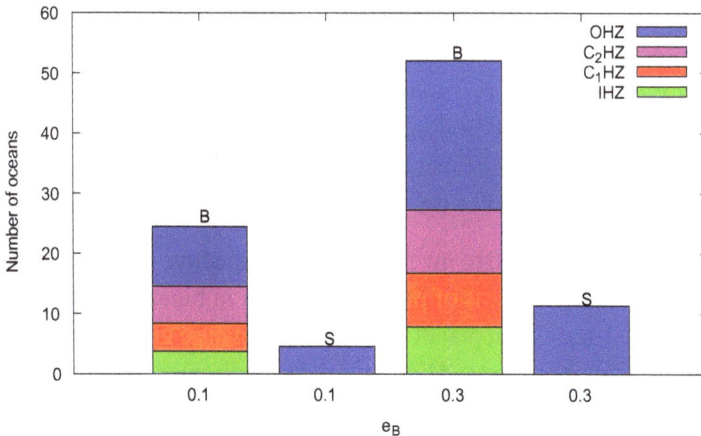

Figure 4.12: Comparison of the water transport in single star (S) and binary star (B) systems. This panel shows the study of two values of e_B as indicated on the x-axis. The color code refers to different sub-HZs (see text).

the respective single-star–planet system (indicated by the same e_B) shows that in both cases a significantly higher amount of water (4–6 times) is transported into the HZ for the binary configurations. Moreover, the icy asteroids cross the HZ in all four sub-HZs, while in single stars the water delivery is only into the OHZ (which extends from 1.49 au to 1.67 au). However, it can be expected that in wider binary star systems, the difference in the results for binary and single stars will be smaller. This study shows clearly that in simple single-star–giant planet systems — which are quite numerous according to the detections[9] — a terrestrial planet at 1 au will probably not have enough water. The reason is that the water transport by icy planetesimals is not efficient enough and, in addition, icy asteroids deliver the water mostly to the outer region of the HZ. In contrast, in binary star systems a larger amount of water is transported into the entire HZ on a shorter time-scale due to stronger perturbations.

4.4 Summary

Numerical simulations of the early stage of planetary formation in binary star systems indicate that we are still facing many open questions regarding where and how planets can grow in such environments. There are many parameters that affect the outcome of the simulations, such as

  (i) the smoothing parameter regarding the disc–protoplanet and protoplanet–protoplanet interaction,
 (ii) the type of boundary conditions (reflecting, outflow, non-reflecting) of the grid in the hydrodynamical part,
(iii) different flux-limiter functions in the advection part of the code, and
 (iv) the orbital evolution of the secondary star.

Future studies are needed to investigate their respective influences and shed more light on the problems of planet formation in binary star systems.

In contrast, terrestrial planet formation using pure gravitational N-body simulations for embryo-sized bodies can be easily performed and result in one or several small planets depending on the initially binary-star–planet configuration. If the binary star hosts also a giant planet, MMRs with this

---

[9]https://exoplanet.eu.

planet play an important role as they restrict the region for the late stage formation and limit the eccentricity of the terrestrial planets to smaller values than in systems without giant planet.

Even if planets can easily form in circumstellar HZs, this does not guarantee that such planets will have habitable environments. The amount of water on a planet's surface seems to be crucial to sustain a temperate environment (Kasting *et al.*, 1993). Therefore, water transport studies are of great importance. For the Earth, two mechanisms seem to be important: (i) endogenous outgassing of primitive material, and (ii) exogenous impact by asteroids and comets. Since neither mechanism can explain the amount of water and the isotope composition of Earth's oceans by itself, models that favour a combination of both sources seem to be more appropriate (Izidoro *et al.*, 2013). The amount of primordial water that is collected during the formation of planets in S-type orbits in binary star systems can be between 4 and 40 Earth oceans, according to the study by Haghighipour and Raymond (2007). They also showed that a higher eccentricity of the binary leads to fewer and dryer terrestrial planets. In such a case, a late veneer water transport will be needed to provide habitable conditions on the planet. In this context, Bancelin *et al.* (2015, 2016) showed that in binary systems the water transport into the HZ is more efficient than in single star systems — the amount of water (in Earth oceans) can be 4–5 times higher.

In any case, the architecture of a binary–planet system is important since perturbations such as MMRs and SRs might influence planet formation and the water transport. The various studies suggest that (nearly) circular motion in the HZ and sufficient perturbations in the planetesimal belt provide better conditions to create a habitable environment for a planet in the HZ.

Chapter 5

Implications of Stellar Binarity

In this chapter, we focus on particular consequences for the habitability due to stellar binarity. The strong influence of the host-star's physical properties on the habitability of a planet is quite obvious. To what extent the stellar output influences the habitability of a planet is, therefore, in the spotlight of current research. Recently, Güdel *et al.* (2014) summarized the astrophysical conditions required for the formation of habitable planetary environments.

In the case of binary stars, a combined influence due to stellar properties and dynamical perturbations can be expected. We, therefore, discuss both the dynamical influence of a secondary star, where we have to distinguish between S- and P-type planetary motion, and specific properties of the stars, where the age, evolution and activity affect the conditions of habitability. In dynamical investigations, the architecture of the binary-star–planet system is of prime importance, since the mutual distances of the celestial bodies and their eccentricities determine the locations of gravitational perturbations. For S-type motion, a faraway secondary star will not have a strong radiative influence on a planet moving in the HZ; for P-type motion, the combined stellar radiation has a stronger influence. Especially the occurrence of wind–wind interacting zones could have an impact on the habitability of planets when frequently crossing these regions.

In the following sections, we discuss for both, S- and P-types, the specific influence of a secondary star where we take into account the recent studies by Pilat-Lohinger *et al.* (2016) and Bazsó *et al.* (2017) for S-type motion, and by Johnstone *et al.* (2015c) for P-type motion. In addition, we mention specific stellar properties that are important for habitable environments.

5.1 S-types: Dynamical Influence of a Secondary Star

To analyze the influence of the secondary star from a dynamical point of view, one has to take into account the constraints of stability in binary stars discussed already in Chapter 2. Orbital calculations of planets in binary star systems have shown that initially circular planetary motion will not remain circular, as it would be the case for a single star. A distant secondary star might cause periodic variations of the planetary eccentricity depending on the parameters of the binary star. In Chapter 6, one can see that these variations in eccentricity affect also the insolation on a planet in the HZ. A comparison of the top and middle panels of Figure 6.3 shows a correlation of both signals. This leads to various definitions of HZs in binary star systems: permanent, extended and averaged HZ (see Eggl *et al.*, 2012). The change in insolation is mainly due to the eccentric motion of the planet which causes periodic approaches to the host-star. The radiative effect of the secondary star on the HZ is discussed in Section 6.3 where we recognize a strong dependence on the stellar distance and mass.

Focusing on the gravitational influence of the secondary star, we consider binary-star–planet configurations consisting of a gas giant and a terrestrial planet where the latter moves in the HZ. The interplay of these components in such an hierarchical four-body problem causes perturbations like mean motion resonances (MMRs) and secular resonances (SRs) whose locations strongly depend on the architecture of the system. By varying the orbital parameters of a system one can analyze the influence of the secondary star. This is visualized when comparing the two panels of Figure 5.1. The figure shows the dynamical behavior of fictitious test-planets in orbits between 0.15 and 1.3 au. In the left panel these test-planets are perturbed only by a gas planet at 1.64 au, while the right panel shows the additional perturbations by a secondary star at 20 au. The binary-star–planet configuration used for this study matches the HD 41004 AB binary system (see Table 7.1).

The left panel of Figure 5.1 shows the dynamics in the restricted three-body problem where only MMRs with the gas giant (vertical structures)

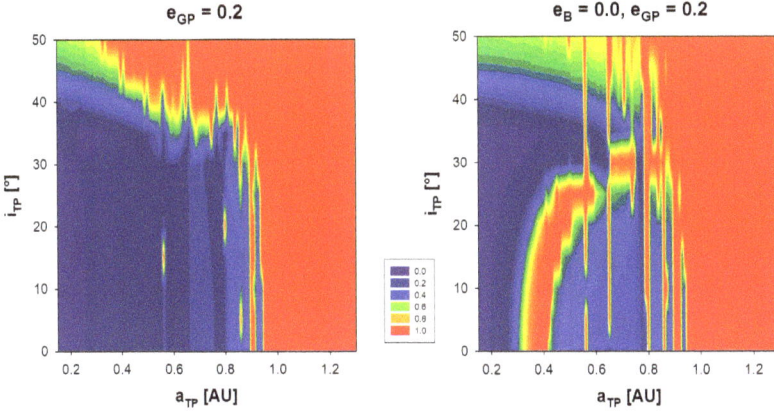

Figure 5.1: Left panel: Maximum eccentricity plot for test-planets perturbed only by a giant planet at 1.64 au. Right panel: Test-planets perturbed by a giant planet at 1.64 au and a secondary star (M-type) at 20 au. The color scale (from 0 to 1) indicates different maximum eccentricities achieved by the test-planets over the whole computation time. Stable motion is labeled by dark-blue areas and red indicates unstable motion.

can be recognized. For the stable region, a cut-off for inclinations $i > 38°$ due to the Kozai resonance (Kozai, 1962) is clearly visible. Once we add a secondary star at 20 au to the system, perturbations at MMRs are stronger and better visible, and an SR (red arched band) arises and causes additional instabilities in the stable area (see right panel). The origin of the SR lies in the behavior of the giant planet. It experiences periodic perturbations at every periastron passage of the two stars, which act on the planet's eccentricity (see the yellow line in Figure 5.2) and argument of perihelion. These perturbations cause a precession of the giant planet's perihelion with the same frequency as that of the test-planets in the region of the red arched band in the right panel of Figure 5.1. Since both panels of Figure 5.1 show about the same extension of the stable area (i.e., blue region) in a_{TP}, we can assume that the parameters a_{GP} and e_{GP} of the giant planet determine the border between stable and chaotic motion. Therefore, the main differences of circumstellar motion in a binary star system compared to that around a single star are the following:

(i) Stronger perturbations at MMRs with the giant planet (which is perturbed by the secondary star).

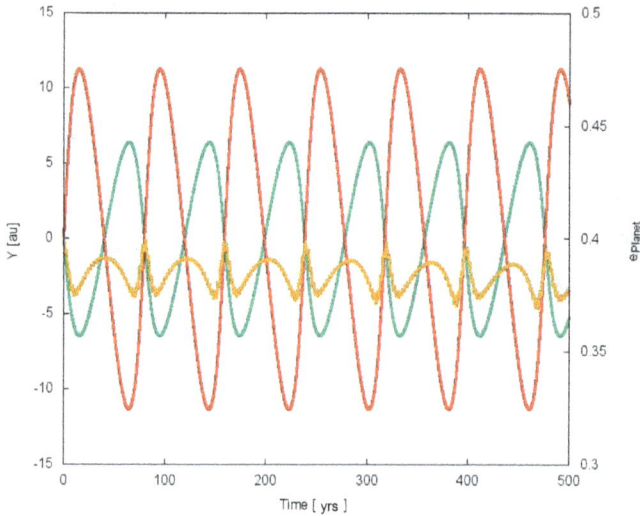

Figure 5.2: Time evolution of the barycentric y-coordinate of the two stars (red and green lines) and of the giant planet's eccentricity (yellow line). The latter indicates jumps at every pericenter passage of the stars which can be seen at multiples of the orbital period of 80 years.

(ii) An SR might appear depending on the binary-star–planet configuration.

(iii) Unstable motion is shifted to higher inclinations ($i_{TP} > 60°$) of the test-planets.

Moreover, the right panel of Figure 5.1 indicates a region of *robust stability*. This is the area between the host-star and the SR that is obviously unperturbed by the giant planet, because the motion of the test-planets remains nearly circular during the entire computation time. This area provides perfect conditions for *dynamical habitability*. Assuming that the HZ of the system lies in the robustly stable region, a planet in the HZ would then always move on a nearly circular orbit without ever leaving the HZ.

5.1.1   *The role of the secondary's semi-major axis*

While the positions of MMRs depend on the giant planet only, the SR is also influenced by the parameters of the secondary star. This is shown in the different panels of Figure 5.3 where the influence of the stellar separation,

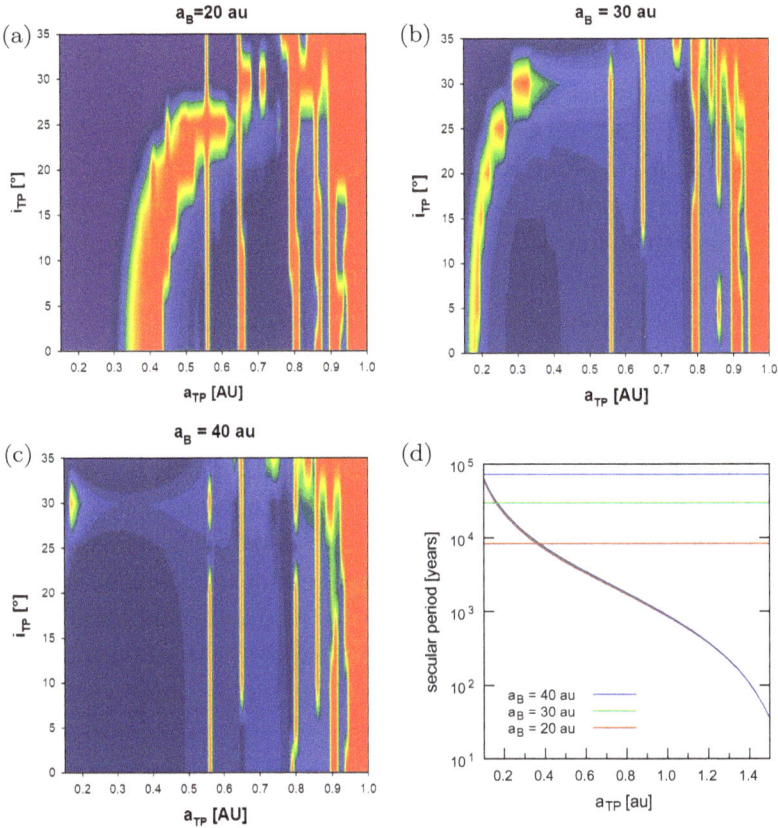

Figure 5.3: (a)–(c): Maximum eccentricity maps for different distances of the two stars as indicated by the title of each panel. The color code is the same as in Figure 5.1 and indicates the different maximum eccentricities achieved by the test-planets over the entire computation time. Red labels the unstable area and dark-blue indicates nearly circular motion. Panel (d) shows the location of the SR calculated with the semi-analytical method described in Chapter 3.

a_B, can be seen from the differences between the 20 to 40 au cases (panels (a–c)). The eccentricities of the binary and the giant planet are initially set to 0.2 in all plots. A comparison of the three maximum eccentricity plots indicates significant variations for the SR when increasing a_B. More precisely, one can recognize a shift of the SR towards the host-star. For $a_B = 30$ au (panel Figure 5.3(b)), the arched red band is still visible but significantly reduced in width and the orbits in this area seem to be stable, but with strong periodic variations in eccentricity (up to $e_{TP} = 0.8$). A further increase of the stellar distance causes again a shift of the SR towards the host-star so that it moves out of the area that we are studying. Only the yellow spot at $i_{TP} = 30°$ reveals the existence of this phenomenon (see panel Figure 5.3(c)). An SR close to the host-star could cause periodic variations in eccentricity for close-in planets that are in general on nearly circular orbits due to tidal forces. The influence of tidal effects is not included here, but it could be an interesting aspect for future studies.

Figure 5.3(d) summarizes the application of our semi-analytical method (see Chapter 3) which helps to compute the location of an SR in coplanar binary-star–planet systems. The different colors represent the result of this method for each of the three configurations of the maximum eccentricity study in Figure 5.3(a–c). One can see that the analytically derived periods of the test-planets between 0.1 and 1.5 au are quite similar in all systems; only the proper period of the gas giant (represented by the horizontal lines) changes significantly when modifying a_B. The intersection of an horizontal line with the curve of the test-planets' proper period gives the location of the SR in a system. These intersections show the same shift of the SR towards the host-star when increasing a_B so that both results were found to be in good agreement.

5.1.2 *The role of the eccentricities of the secondary and the planet*

When setting a_B to 20 au for example, and varying the eccentricities e_B and e_{GP}, as shown in the panels (a–d) of Figure 5.4, we find that the planet's eccentricity affects the dynamics in the studied area much more than the binary's eccentricity. An increase of e_{GP} decreases the stable zone significantly, which can be seen by comparing Figure 5.4(a) and (b). Especially the area around the SR and towards the giant planet is severely perturbed if e_{GP} is increased from 0.2 to 0.4. An increase in e_B from 0.2 to 0.4 causes only a slight shift of the SR to the right (compare Figure 5.4(a) and (c)).

The application of our semi-analytical method to these binary-star–planet configurations confirms these findings (see Figure 5.4(d)). The intersections of the horizontal lines with the black curve representing the analytical solution for the proper periods of the test-planets show also a

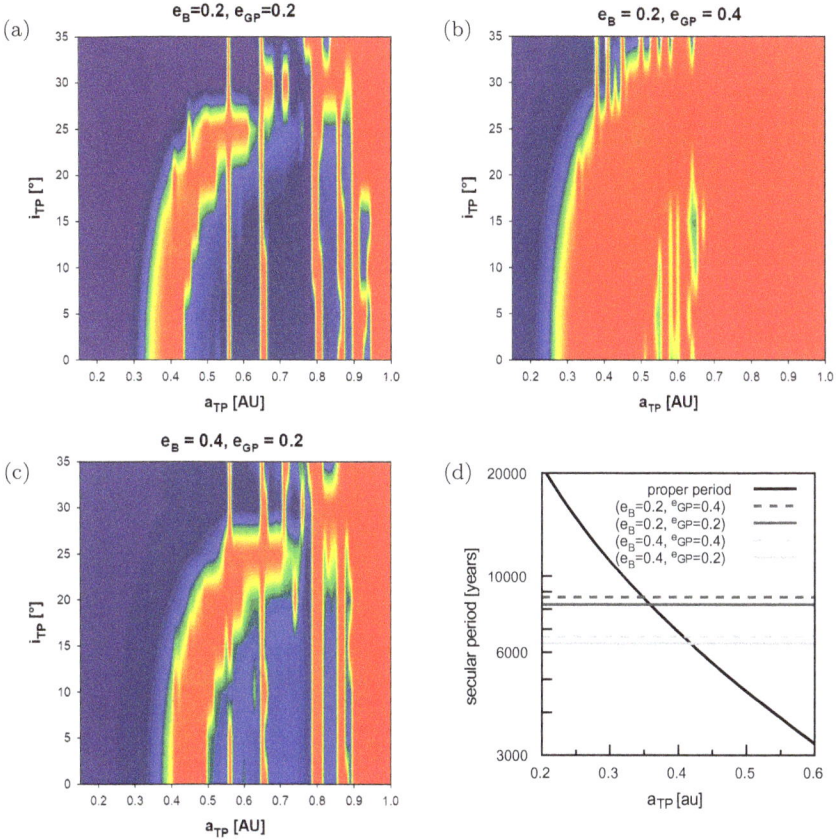

Figure 5.4: (a)–(c) Maximum eccentricity plots for planetary motion in a binary star system with a separation of 20 au. The eccentricities e_B and e_P are either 0.2 or 0.4 as indicated in the title. The color code is the same as in Figure 5.1. Panel (d) shows the result of different configurations (as indicated in the legend) when applying our semi-analytical method. The black curve represents proper periods of the test-planets which were determined analytically. The horizontal lines show the proper periods of the giant planet in the various binary configurations. The location of an SR (a_{SR}) is at the intersection of an horizontal line with the black curve.

strong shift of the SR for a change in e_B from 0.2 to 0.4. The different grey shades indicate different eccentricities of the binary: dark grey lines for $e_B = 0.2$, and light grey lines for $e_B = 0.4$. Different eccentricities of the giant planet are shown by different line styles: full lines are for $e_{GP} = 0.2$, while the dashed lines are for $e_{GP} = 0.4$. A comparison of the proper

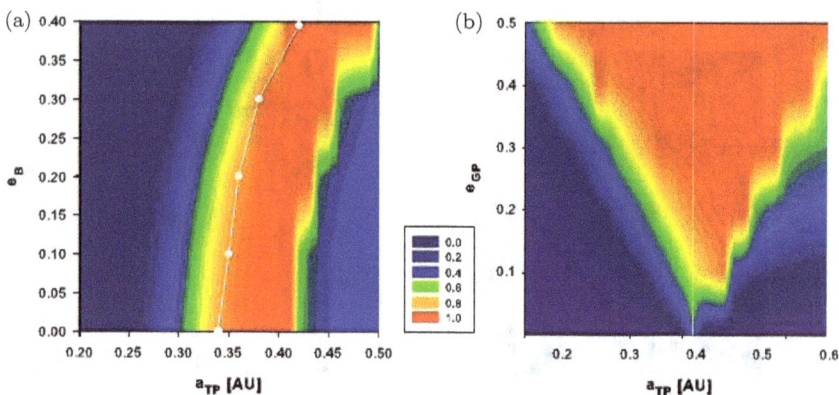

Figure 5.5: Maximum eccentricity maps for binary-star–planet configurations with $a_B = 20$ au. Panel (a) shows the displacement of the perturbed area (red) for a fixed value of $e_{GP} = 0.2$ when e_B (y-axis) is increased. Panel (b) shows an enlargement of the perturbed area (red) for a fixed $e_B = 0.2$ when e_{GP} (y-axis) is increased.

periods for two eccentricities of the giant planet and a certain e_B shows only a small shift of the SR location.

Of course, not only the position of an SR is important; the extension of the perturbed area also plays a role and can vary a lot depending on e_{GP} (see Figure 5.5, panel (b)). This figure shows that the width of the SR is mainly a function of e_{GP} (panel (b)) and confirms that changes in e_B will shift the location of the SR while an increase of e_{GP} enlarges the perturbed area (red) which is visualized by a V-shape[1] for the perturbation. When the motion of the giant planet is circular (i.e., $e_{GP} = 0$), the SR is confined to a very small range of orbital distances, as can be seen by the thin green spike; here the eccentricity has variations up to 0.6 but no escapes occur. For escape orbits within the SR, the eccentricity e_{GP} has to be larger than 0.1. The computed location of the SR (see the green spike, which coincides with the value determined by our semi-analytical method) is not in the center of the resonant region. More precisely, it defines the inner edge of the SR where the maximum eccentricities are larger than the initial e_{GP}.

---

[1] The same shape is known for MMRs in such semi-major axis–eccentricity maps.

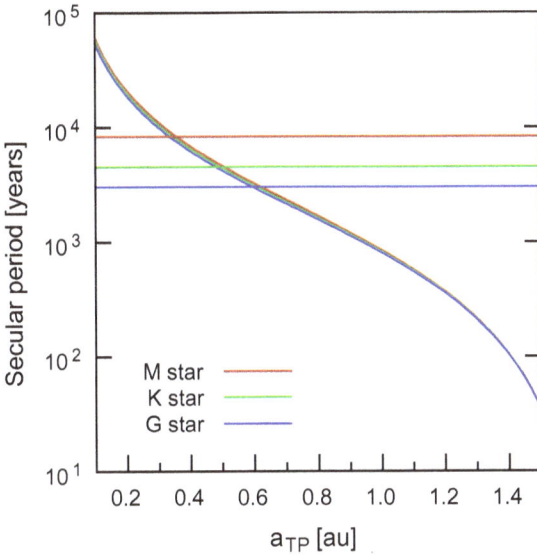

Figure 5.6: Locations of the SR for different binary-star–planet configurations when the giant planet orbits a G-type star at 1.64 au and the secondary star is at 20 au. The mass of the secondary corresponds to either M-/K-/G-type stars of 0.4/0.7/1.0 solar masses, respectively. The location of the SR is the intersection of the giant planet's proper period (horizontal lines) and the analytically determined proper periods of the test-planets (curves). The figure shows a strong variation in the location of the SR for various masses of the secondary star.

5.1.3   *The role of the secondary star's mass*

The influence of the mass of the secondary star is displayed in Figure 5.6, which shows different locations of the SR for different stellar types of the secondary star. For low-mass stars, such as an M-type star with a mass of 0.4 $M_\odot$, the SR is closer to the host-star than for a more massive secondary, such as a K-type star (e.g., with a mass of 0.7 $M_\odot$) or a G-type star (e.g., with a mass of 1 $M_\odot$). Using the semi-analytical method from Chapter 3 to determine the location of the SR, we find nearly similar proper periods for the test-planets (see the curves in Figure 5.6), but different values for the giant planet at 1.64 au (horizontal lines in that figure). The shift of the intersections of the horizontal lines with the respective curve (of the same color) shows a displacement in the direction away from the host-star for an increasing mass of the secondary.

Summarizing these numerical results and taking into account the theory described in Chapter 3, it is obvious that variations of the mass, the semi-major axis, and the eccentricity of the secondary star and/or the giant planet cause also changes of the proper frequencies of these bodies, which modifies then the location of the SR. Therefore we can conclude that (see also Pilat-Lohinger *et al.*, 2016)

(i) an increase of the separation of the two stars shifts the SR towards the host-star,
(ii) increasing the secondary's mass will move the location of the SR towards the secondary,
(iii) the eccentricity of the binary influences the location of the SR,
(iv) and the eccentricity of the giant planet is responsible for the width of the SR.

5.2 P-types: Dynamical Influence of a Secondary Star

Dynamical considerations for habitability studies of planets in P-type motion around both stars show also a slight influence on the planet's eccentricity which is varying in a similar way as it does for S-type motion. However, Figure 6.3 (middle panels) indicates a significantly smaller amplitude of e_P for P-type orbits where the maximum eccentricity is <0.06 even for high eccentricities of the binary (e.g., $e_B = 0.5$, which is represented by the red curve in the right panel). Similar values were found for the dynamics by Johnstone *et al.* (2015c) for the permanent HZ (PHZ) where a G2V-G2V type binary with $a_B = 0.5$ au has been considered.

As an example, we show in Figure 5.7 the HZ in this binary for e_B from 0 to 0.9 — even if we do not expect highly eccentric motion for such tight binaries. This figure shows the different HZs[2] defined by the method of Eggl *et al.* (2012): blue denotes the PHZ (1.65–2.27 au), green the EHZ (1.40–2.36 au) and yellow the AHZ (1.37–2.39 au). The solid and dotted vertical black lines indicate the inner and outer limits of the classical HZ calculated using the method of Kopparapu *et al.* (2014) for single stars,

---

[2]PHZ = permanent HZ, EHZ = extended HZ and AHZ = averaged HZ; for details see Chapter 6.

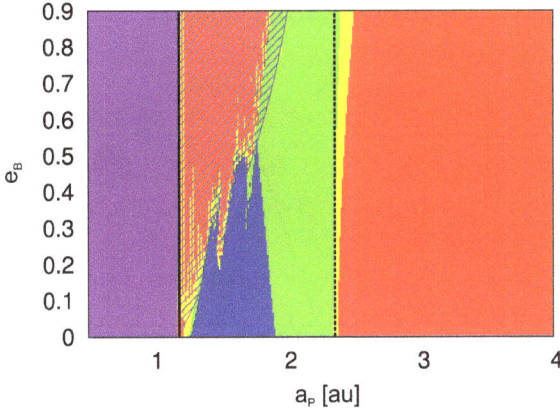

Figure 5.7: Location of the HZ (x-axis) in a G2V-G2V binary star system for different eccentricities of the binary (y-axis). We distinguish between stable (red area) and chaotic (purple area) motion. For the HZ, blue indicates the permanent HZ, green the extended HZ and yellow the averaged HZ (as defined in Chapter 6). To show details of the HZ for higher e_B, the border of chaotic motion within the HZ is displayed by the hatched area. The solid and dotted vertical black lines show the inner and outer limits of the classical HZ calculated using the method of Kopparapu *et al.* (2014) for single stars, assuming that a central star's bolometric luminosity is equal to the sum of the bolometric luminosities of the two stars in the binary system.

assuming that the central star's bolometric luminosity equals to the sums of the bolometric luminosities of the two stars in the binary system. For $e_B = 0$, the borders of the classical HZ are quite similar to those of the AHZ, where higher variations in eccentricity are possible. A study of the maximum eccentricity of test-planets in the area of the classical HZ of this system shows that maximum eccentricities up to 0.1 can arise for test-planets with $a_P \leq 1.4$ au. In the outer region of the HZ ($a_P > 2$ au) the test-planets show only minor variation in eccentricity (<0.04). From this it follows that the concept of various HZs introduced by Eggl *et al.* (2012) is essentially important for the inner border of the HZ.

5.3 Specific Stellar Influence

The HZ is — according to its definition — primarily determined by the luminosity and spectral properties of the host-star (see Chapter 6); consequently, the stellar type plays a major role in studies of habitability.

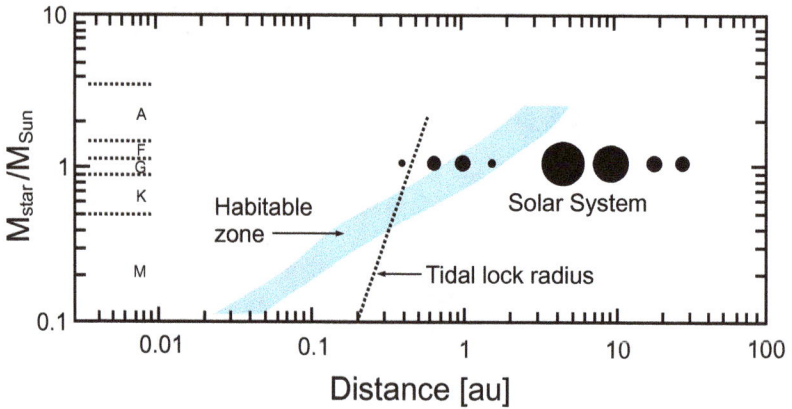

Figure 5.8: Location of the HZ (blue area) for main sequence stars from A to M type.

Kasting *et al.* (1993) presented the first investigation that considered the concept of HZs for different stellar types, including A to M main sequence stars. Due to the luminosity of these stars, a bright A class star has its HZ at a larger distance, and for a faint M star the HZ moves quite close to the host-star, as illustrated in Figure 5.8. This figure indicates in addition, that planetary motion in the HZ of low mass stars will be tidally locked when they are close to their host-star. The tidal locking applies for M- and K-type stars.

In Figure 5.8 one can see the "water based HZ" (blue band) for the different main-sequence stars where a planet moving in this area has the ability that water is liquid on the planet's surface. This, however, requires that the planet is massive enough[3] and has a relatively dense atmosphere. Moreover, the evolution of the planetary atmosphere is strongly related to the evolution of the host-star.

5.3.1 *Stellar type: Age, evolution and activity*

At present, the solar activity is not likely to change the conditions for Earth's habitability. However, the early Sun was more active, so the solar X-ray and

---

[3] Mars in the Solar System is in the HZ, but is not massive enough to fulfill the requirement of liquid water on its surface.

EUV emission (together XUV), solar flares, coronal mass ejections (CMEs) and the solar wind (see e.g., Johnstone *et al.* (2015a,b) or Tu *et al.* (2015)) had stronger effects on the evolution of early Earth (see e.g., Hanslmeier, 2007). A recent study by Tu *et al.* (2015) showed that the young Sun could have had different rotational evolution tracks ranging from slow to fast rotators. The rotation is important as it affects the activity of a star regarding the magnetic field, the stellar wind and the high energy radiation (i.e., the extreme ultraviolet and the X-rays). This uncertainty in the early phase of a Sun-like star extends to an age of 500 million years of the stellar evolution for G stars. During this period, planet formation takes place. Therefore, the rotation of the host-star might also be crucial for the evolution of a planetary system.

The planetary atmosphere is especially affected by the high-energy radiation of a young star for whose evolution is not fully understood today. Intense studies of our Sun suggest a lower photospheric luminosity (of about 30% less than today) when it arrived on the main-sequence (see e.g., Güdel, 2007). During its time on the main-sequence, our Sun's magnetic activity declined. The decrease of activity with age depends on the stellar mass. Studies of stellar evolution show, especially for A and F class stars, limited habitability due to their short lifetime on the main-sequence (see Figure 14 in Kasting *et al.*, 1993). Low mass K and M stars, in contrast, have nearly constant HZ borders for about 10^{10} years. However, these low mass stars are known to emit high levels of short-wavelength radiation (X-ray, EUV, far UV and UV) and their activity phase is significantly longer than for more massive stars. In this phase, the flaring activity, stellar winds and CMEs are enhanced and might therefore influence the conditions of habitability of a terrestrial planet in the HZ. Moreover, the fact that the HZs of M-type stars are much closer to the star (see Figure 5.8) might intensify these perturbations that can heat the upper planetary atmosphere leading to an enhanced rate of atmospheric loss. The winds and CMEs affect the planet's magnetosphere leading to an atmospheric erosion (Khodachenko *et al.*, 2007). Various studies of M-class main-sequence stars revealed a duration for the high activity phase of up to 10^9 years depending on the stellar mass (Lammer, 2007; Lammer *et al.*, 2007, 2011; Lichtenegger *et al.*, 2010). This behavior is displayed in Figure 5.9 where the red lines indicate the evolution of various M-class stars with masses from 0.1 to 0.6 $M_\odot$.

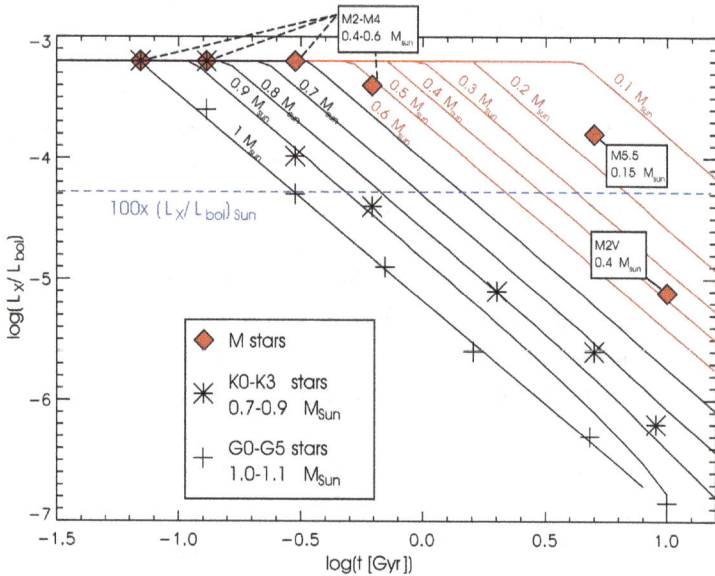

Figure 5.9: Evolution of stellar radiation for low-mass main-sequence stars.

They all show an enhanced XUV flux (a hundred times that of the Sun) for a very long time as indicated by the horizontal dashed blue line. This behavior certainly restricts the conditions for habitability.

5.3.2 *Stellar winds*

Johnstone *et al.* (2015c) studied the wind conditions that circumbinary habitable planets might experience when orbiting a binary star system consisting of two G-type stars. Their wind interaction model[4] yields a wind–wind collision region with enhanced density and temperature, separated from the winds of the individual stars by two strong shocks as indicated by Figure 5.10.

A potentially habitable planet would have to pass through these shocks four times per orbit in a rotating frame of reference. This corresponds to 28 times per orbit in the inertial frame of reference, such that a planet

---

[4]They used the results of a 1D model for the slow solar wind as input to a 3D hydrodynamical wind interaction model, including Coriolis forces due to the orbital motions of the two stars.

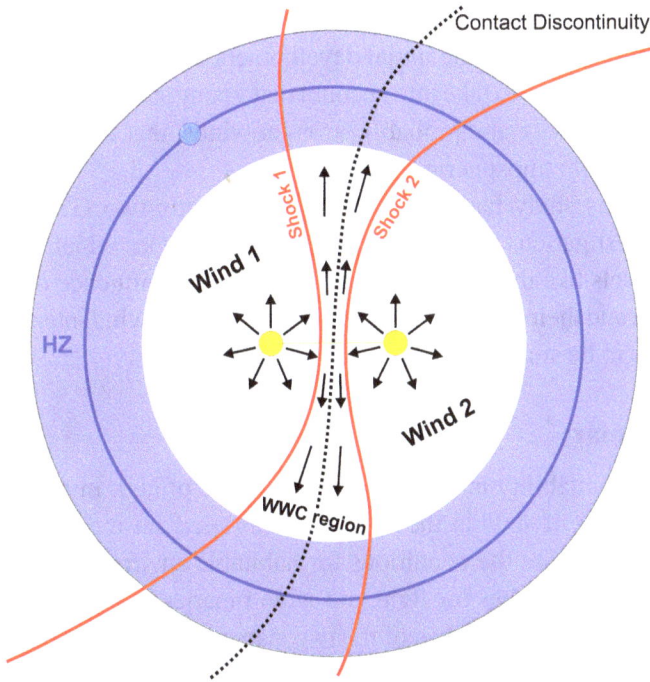

Figure 5.10: Main features of wind–wind interactions in a binary system with two identical winds. The wind–wind collision region (WWC) takes the form of two shock waves (red lines) and has a spiral geometry due to the orbital motion of the stars. The blue ring indicates the HZ which is obviously influenced by the WWC region.

would spend almost a quarter of each orbital period in regions of enhanced wind density and temperature. It is well known that stellar winds may influence the evolution of a planet in the HZ as these kinds of output can ionize and expand planetary atmospheres and cause, therefore, the loss of atmospheric particles (see Watson *et al.*, 1981; Lammer *et al.*, 2003, 2013; Tian *et al.*, 2005, 2008; Khodachenko *et al.*, 2012; Kislyakova *et al.*, 2014). However, the study by Johnstone *et al.* (2015c) does not assume severe influences on a planet like the current Earth that could cause an increased atmospheric loss due to interactions with the wind. But for a much thicker hydrogen dominated atmosphere — that could have existed on the primordial Earth according to Hayashi *et al.* (1979), Sekiya *et al.* (1988) and Lammer *et al.* (2014) — the magnetospheric compression due to the interacting winds could lead to additional stripping of the outer layers of the planetary atmosphere. In that case, the binary star would have a significant influence on the initial development of a planetary atmosphere. In addition, we have to take into account that young stars show higher levels of magnetic activity and probably stronger winds, that could lead to even higher levels of atmospheric loss.

Since this study has been carried out for only two G2V type stars, further investigations are needed for other stellar types which have higher activity levels like the M-class stars. Moreover, the influence of the stellar separation and their orbital eccentricities on the wind–wind interaction zone also needs to be studied.

5.4 Summary

Studies of habitability in binary star systems are of high importance since a large number of stars in the solar neighborhood are members of stellar systems. To explore the conditions for habitable environments, there will be different approaches for tight and wide binaries. For tight double star systems, the stellar properties of the two stars play a major role, especially when their distance is small, so that stellar activities (like CMEs, winds or XUV flux) of both stars will intersect and significantly modify the environment conditions for a planet. In wide binary stars, the dynamical perturbations of a faraway secondary star are more effective and have to be considered for studies of habitability, which may lead to a combined dynamical-radiative influence for the HZs in such systems.

Chapter 6

Habitable Zones in Binary Star Systems

Given the effort and observational resources required to detect Earth-like planets, especially in binary star systems (e.g., Endl *et al.*, 2015), it is desirable to identify and prioritize candidate systems that are, at least in theory, capable of hosting habitable worlds. Habitable zones (HZs) as suggested by Kasting *et al.* (1993) are a potent tool to achieve such a categorization. Many Sun-like stars are part of multiple systems. One may wonder how HZs in binary star systems can be calculated and how similar they are to HZs around single stars. Following a brief review of some of the key results for HZs around single stars, we explore HZs in binary star systems that are based on insolation geometry. Constraints on the existence of planets in circumbinary habitable zones (CBHZs) and circumstellar habitable zones (CSHZs) imposed by orbital dynamics are discussed. The consequences of orbital dynamics regarding insolation are then presented and dynamically informed HZs are introduced. A brief discussion on self-consistent climate simulations of planets in binary star systems and the potential superhabitability of planets in binary star systems concludes this chapter.

6.1 Habitable Zone History

Huang (1959) was among the first to coin the term "habitable zone". To him, a HZ was the region around a star where:

> "The heat received by the living beings on a planet must be neither too large nor too small."

Huang assumed the amount of star-light arriving at the top of a planet's atmosphere (insolation) was the main driver for a planet's climate and surface temperature. Planets orbiting other types of stars would be habitable if they received insolation quantities similar to the Earth in the Solar System:

$$\frac{l_\star}{4\pi r^2} = \frac{l_\odot}{4\pi r_\oplus^2} = s_\odot. \tag{6.1}$$

In Equation (6.1), $l_\star$ is the luminosity of the star, $l_\odot$ is the luminosity of the Sun, and r is the distance of the Earth-like planet on its circular orbit around its host star. The quantity s is the amount of energy per square meter per second arriving at the planet and $r_\oplus$ denotes the mean distance from the Earth to the Sun. Currently, the Earth receives approximately $s_\odot \approx 1361$ W m^{-2}, which is called the "solar constant". Equation (6.1) states that a Sun-like star casts one solar constant's worth of radiation on a planet that is on a circular orbit with a semi-major axis of one astronomical unit. To avoid the explicit numerical factors occurring in Equation (6.1), we can use a more adequate set of units. With $L = l_\star/l_\odot$. [au]^2, $S = s/s_\odot$ and r given in astronomical units, the corresponding equations are

$$\frac{L}{r^2} = S, \quad r = \sqrt{\frac{L}{S}}. \tag{6.2}$$

The Earth itself is not on a perfectly circular orbit. Together with small variations in the Sun's luminosity, the Earth's variable distance to the Sun causes changes in the amount of light our planet receives. Since the Earth is still habitable, it is not unreasonable to think that the Earth's climate remains robust against collapse for a range of higher (S_I) and lower (S_O)

insolation values. Huang (1959, 1960) proposed that a terrestrial planet could remain habitable for insolation values between $S_I = 5$ and $S_O = 0.1$, corresponding to five times the solar constant and a tenth of a solar constant, respectively. Using those insolation values as limits, Huang constructed HZs with inner (r_I) and outer (r_O) borders

$$r_I = \sqrt{\frac{L}{S_I}}, \quad r_O = \sqrt{\frac{L}{S_O}}. \tag{6.3}$$

Huang's insolation limits would result in a HZ stretching from well inside Venus' orbit all the way to the asteroid belt. In retrospect, Huang's choices for how much radiation makes a planet have "just the right temperature" seem somewhat arbitrary. A more tangible approach to defining habitability limits is based on determining where a planet can sustain liquid water on its surface.

6.2 Spectral Weights

The presence of considerable amounts of liquid water on a planet's surface plays a crucial role in regulating climates (e.g., Rasool and de Bergh, 1970; Hart, 1978, 1979; Kasting, 1988; Kasting *et al.*, 1993; Selsis *et al.*, 2007; Kopparapu *et al.*, 2013; Leconte *et al.*, 2013; Godolt *et al.*, 2016; Popp *et al.*, 2016). However, water can only fulfill its role as a climate-buffer in a limited temperature domain (Kasting *et al.*, 1993). If the energy input into the climate system exceeds certain limits, the system enters a "runaway state".

Runaway states are unstable conditions in which small changes in insolation are no longer causing similarly small variations in a planet's climate. Such processes generally lead to extreme surface temperatures (e.g., Spiegel *et al.*, 2008). Known runaway states include a total evaporation of surface oceans, which would lead to photo-dissociation of water molecules in the upper atmosphere and, ultimately, cause a severe loss of hydrogen to space. A cold trap runaway process, on the other hand, starts with a freeze-out of the most effective greenhouse gases in the atmosphere. This, in turn, results in a drop in planetary surface temperatures facilitating glaciation. The consequent increase in surface albedo causes more light to be reflected to space, cooling the planet further. The final outcome is a "snowball state" where all of the planet is covered in snow and ice.

In order to define HZ borders one can determine insolation thresholds, $S_{I,O}$, that would lead to climatic runaway states on a planet with an atmosphere not too dissimilar to that of the Earth. Once found, those insolation limits can be translated to HZ borders using Equation (6.3).

A key insight regarding runaway limits is that not only the quantity of star-light but also the light's spectral distribution is relevant to planetary climates (Kasting *et al.*, 1993). For stars on the main-sequence, the latter can be approximately described through their effective temperature (T_{eff}), i.e., the energy equivalent black body emission profile that is fit to the actual stellar spectrum. Given the same incident power, Kasting *et al.* (1993) found that the light originating from an M-class star is more potent in heating the Earth compared to the light of an F-class star. In other words, given two stars with the same luminosity, the star with the lower effective temperature (T_{eff}) would have its HZ borders further out. The insolation thresholds $S_{I,O}$ can, thus, be parametrized as a function of T_{eff}. Kopparapu *et al.* (2014) proposed the following fourth-order fit functions:

$$S_I = 1.107 + 1.332 \times 10^{-4}T + 1.580 \times 10^{-8}T^2$$
$$- 8.308 \times 10^{-12}T^3 - 1.931 \times 10^{-15}T^4,$$
$$S_O = 0.356 + 6.171 \times 10^{-5}T + 1.698 \times 10^{-9}T^2$$
$$- 3.198 \times 10^{-12}T^3 - 5.575 \times 10^{-16}T^4,$$

(6.4)

which represent insolation values corresponding to the runaway greenhouse (S_I) and the maximum greenhouse (S_O) atmospheric collapse limits. Here, $T = T_{\text{eff}}/(1\text{K}) - 5780$. Inserting $S_{I,O}$ into Equations (6.3) yields the "classical" HZ borders for main-sequence stars. The insolation thresholds $S_{I,O}$ contain information on how much light of a specific spectral distribution is necessary to trigger runaway states. Hence, they are often referred to as "spectral weights". Calculating HZ borders via spectral weights is practical. However, insolation limits are not cast in stone; they depend on the underlying climate models and solvents (Godolt *et al.*, 2016; Ludwig *et al.*, 2016).

6.3 Double Stars

Almost immediately after investigating HZs around single stars, Huang (1960) published a similar study for double star systems. In order to account for the second star as an additional source of radiation, one can modify Equation (6.2) as follows:

$$\frac{L_A}{r_A^2} + \frac{L_B}{r_B^2} = S. \tag{6.5}$$

Here, L_A and L_B are the luminosities of the stars A and B, r_A and r_B represent the distances between the respective star and the planet in astronomical units, and S denotes the normalized amount of light arriving at the planet. In order to find binary star HZ borders, one can follow the approach leading to Equations (6.3), i.e., choose $S = S_I$ and $S = S_O$ values, and find the corresponding regions around stars A and B where Equation (6.5) holds. By not introducing individual spectral weights for the stars, we have implicitly assumed that the light of both stars has the same spectral properties, i.e., the stars have similar effective temperatures. Strictly speaking, this makes Equation (6.5) only applicable to binary configurations with identical stellar components on the main-sequence.

6.3.1 Spectral weights for double stars

Not all binaries have similar stellar components. We have to account for the fact that stars have distinct luminosities and spectral properties. Kane and Hinkel (2013) suggested that this issue can be solved by calculating the actual spectral distribution of the light arriving at the planet. This is done by superimposing the spectra of both stars weighted with the respective star-to-planet distances. Eggl *et al.* (2012), Haghighipour and Kaltenegger (2013), Kaltenegger and Haghighipour (2013) and Cuntz (2014) have adopted a slightly different approach. Instead of combining the spectra and investigating the impact on the Earth's atmosphere, they assumed that each star heats the Earth independently. Attaching the respective spectral weight to the contribution of each star we can rewrite Equation (6.5):

$$\frac{L_A}{SA_I}\frac{1}{r_A^2} + \frac{L_B}{SB_I}\frac{1}{r_B^2} = 1, \quad \frac{L_A}{SA_O}\frac{1}{r_A^2} + \frac{L_B}{SB_O}\frac{1}{r_B^2} = 1, \tag{6.6}$$

where $SA_{I,O} = S_{I,O}(T_{\text{eff}}(A))$ are the spectral weights for the inner and outer edges of the single star HZ using the effective temperature of Star A. Similarly, $SB_{I,O} = S_{I,O}(T_{\text{eff}}(B))$ represents the spectral weights for the single star HZ borders of Star B. Both the luminosity and the spectral weights are approximately constant for a given star, at least as long as the star is on the main sequence. We can, therefore, introduce the spectrally weighted luminosities

$$\mathbb{A}_{I,O} = L_A/SA_{I,O}, \qquad \mathbb{B}_{I,O} = L_B/SB_{I,O},$$

that allow us to write Equations (6.6) in an even more concise form

$$\frac{\mathbb{A}_{I,O}}{r_A^2} + \frac{\mathbb{B}_{I,O}}{r_B^2} = 1. \tag{6.7}$$

Equation (6.7) constitutes the foundation on which the rest of this chapter is built. Note that in the unit system we have chosen, $[\mathbb{A}] = [\mathbb{B}] = [\text{au}^2]$. Expression (6.7) represents two equations; one is for the inner (I) and the other is for the outer (O) border of the classical HZ. Acknowledging this fact, we shall drop the indices I and O and only state them explicitly when needed.

6.3.2 *Isophote-based and radiative HZs*

How does a second source of radiation influence the HZ, or HZs? Figure 6.1 shows the instantaneous insolation values for two S-type binary star systems (e.g., Kane and Hinkel, 2013). The color code represents the solar constant at each location, i.e., S in Equation (6.5). The system in the top panel is akin to α Centauri, while in the bottom panel α Centauri B was swapped with an M5V class star. Further details on the stellar parameters used to generate those plots are given in Table 6.1.

The continuous black lines in Figure 6.1 trace specific values of insolation (isophotes). The four black curves are generated using spectrally weighted insolation values that correspond to the inner and outer HZ limits for each star. We shall refer to the area around each star enclosed by those isophotes as an "isophote-based habitable zone" (IHZ).

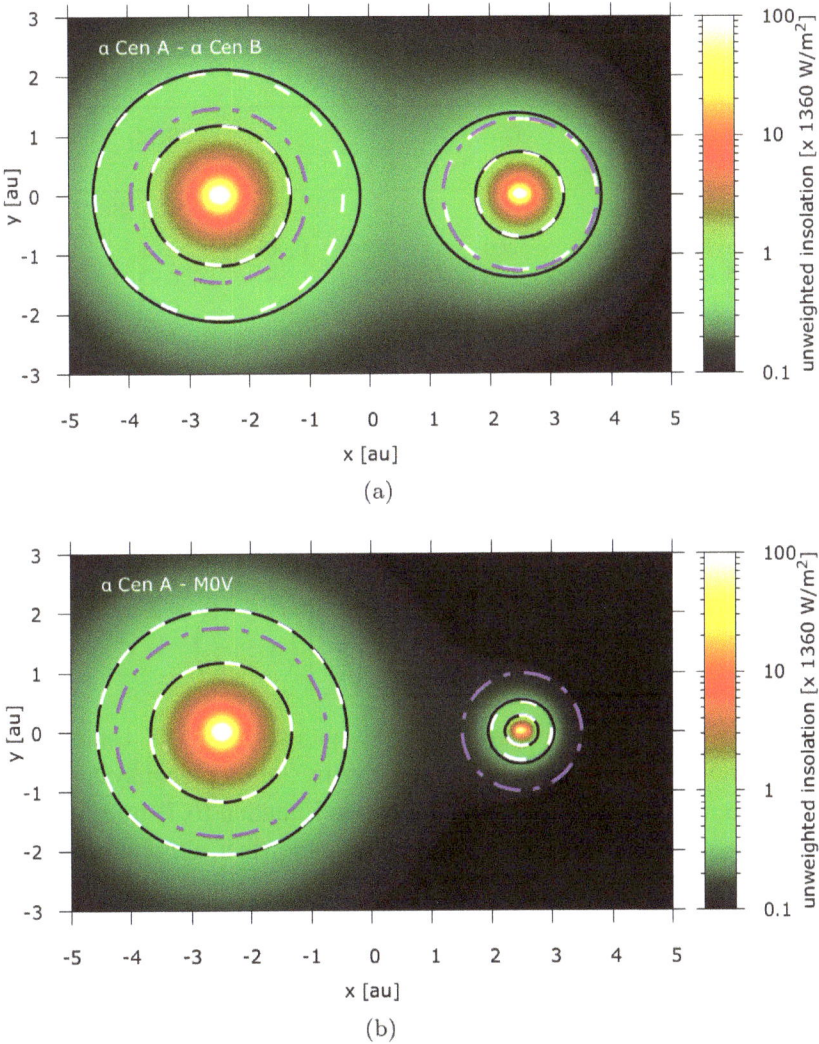

Figure 6.1: Two S-type binary star systems with different stellar components. (a) Shows a system similar to α Centauri with two roughly Sun-like stars, while (b) shows α Centauri A with a less luminous M5V companion. The stars' orbits are circular with a semi-major axis of $a_B = 5$ au. The amount of radiation a planet receives at any point in the system is color-coded. The continuous black lines represent IHZs, which are solutions of Equation (6.7) using spectral weights according to Equations (6.4). The white dashed lines are SSHZs. Planetary orbits inside the purple dashed-dotted circles are dynamically stable.

Table 6.1: Datasheet for main-sequence stars referred to in this chapter (see Welsh *et al.*, 2012; Thévenin *et al.*, 2002; Kervella *et al.*, 2003; Zombeck, 2006). SSHZ$_{I,O}$ symbolizes the inner and outer single star HZ borders, respectively (Kopparapu *et al.*, 2014).

Star	$L/L\odot$	T_{eff} (K)	$R/R\odot$	$M/M\odot$	SSHZ$_I$ (au)	SSHZ$_O$ (au)
α Centauri A	1.52	5790	1.227	1.1	1.17	2.06
α Centauri B	0.50	5260	0.865	0.907	0.71	1.27
Kepler-35 A	0.94	5606	1.03	0.89	0.93	1.65
Kepler-35 B	0.41	5202	0.79	0.81	0.63	1.13
F5V	2.5	6540	1.2	1.3	1.44	2.49
G2V	1	5777	1	1	0.95	1.68
M5V	0.008	3120	0.32	0.21	0.09	0.18

The IHZ borders are contour solutions of Equation (6.7) where we substitute

$$r_A = \left[\left(x + \frac{d}{2} \right)^2 + y^2 + z^2 \right]^{1/2}, \qquad r_B = \left[\left(x - \frac{d}{2} \right)^2 + y^2 + z^2 \right]^{1/2}.$$

Here, x, y and z are Cartesian coordinates with respect to the center of the graph, and d is the distance between the stars. For co-planar systems, i.e., $z = 0$ au, one can find analytic solutions to Equation (6.7) by expressing $y = f(x, d, \mathbb{A}, \mathbb{B})$. This leads to a quartic equation that, although unwieldy, can be solved (Cuntz, 2014). Alternatively, one can use either numerical methods or analytic approximations to solve Equation (6.7).

The dashed white lines in Figure 6.1 represent single star HZ (SSHZ) insolation limits, i.e., solutions of Equation (6.2), where the contribution of the second star is ignored. If a Sun-like star has a very faint companion as shown in Figure 6.1(b), there is little difference between the double star isophotes and the single star isophotes. This indicates that the second star does not contribute much to the primary's HZ in terms of insolation—and vice versa. In contrast, there is a clear difference between the single and double star isophotes in the α Centauri-based system (Figure 6.1(a)). The increased overall flux of the two Sun-like stars causes the IHZs, in particular the outer edges, to extend towards each other. In S-type

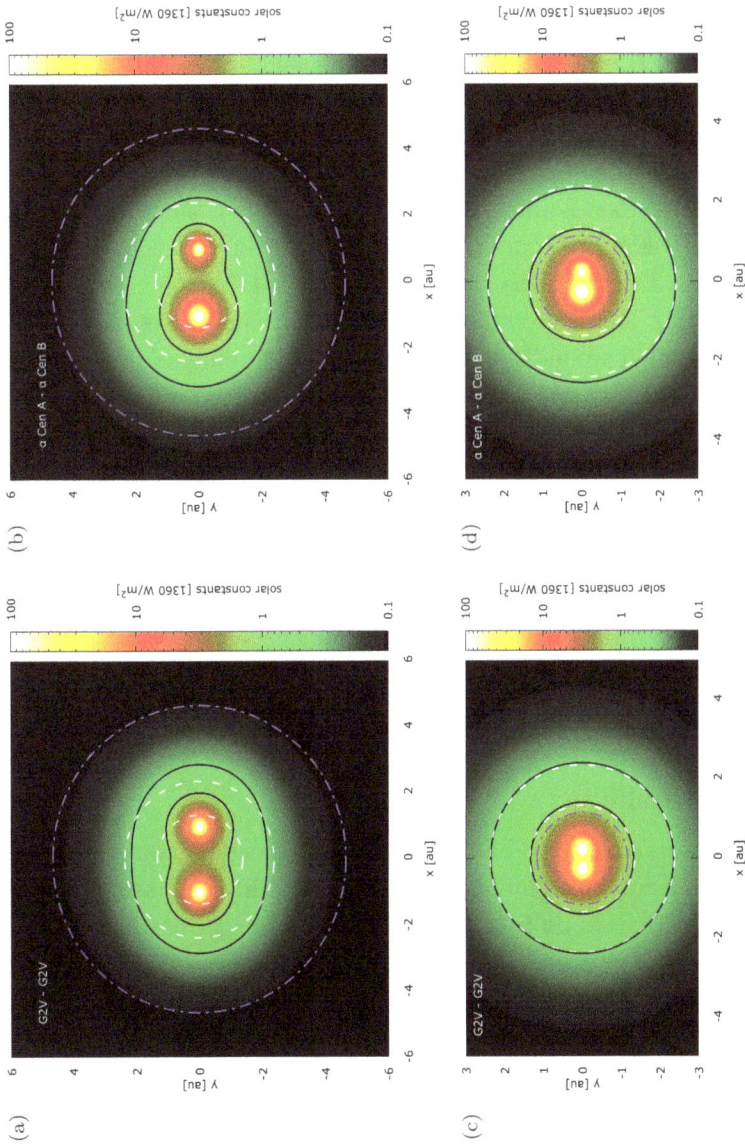

Figure 6.2: Same as Figure 6.1, only for P-type systems. The left column shows two G2V-G2V systems with semi-major axes of (a) $a = 3$ au and (c) $a = 0.5$ au. Panels (b) and (d) show α Centauri-like configurations on orbits similar to those shown in panels (a) and (c), respectively. In contrast to S-type systems, the region beyond the purple line is dynamically stable.

systems, the largest displacement of the isophotes is registered along the line connecting the centers of the two stars. For binary stars on elliptic orbits, the isophote displacement is a function of their mutual distance d and is therefore time-dependent. In order to account for this effect, Müller and Haghighipour (2014) introduced rotating, pulsating IHZs. Having time-varying HZ borders means that IHZs sweep over planets on relatively short timescales. This leads to the problem of determining to which degree planets that are only partly inside HZs are actually habitable — a topic of an ongoing investigation (Williams and Pollard, 2002; Dressing *et al.*, 2010; Bolmont *et al.*, 2016). Circumnavigating those challenges, Cuntz (2014, 2015) introduced the "radiative habitable zone" (RHZ). In an analogy to SSHZs, the RHZ is based on the assumption that planets are moving on circular orbits either around one or both of the stars forming the binary. The RHZ is defined as the largest spherical shell that can be inscribed in the IHZ. The reader is referred to Cuntz (2014, 2015) for the exact expressions.

In very tight binaries, the individual IHZs of both stars merge into a single circumbinary IHZ, as can be seen in Figure 6.2. The merging of separate IHZs into a single circumbinary HZ has a profound impact on RHZs. In contrast to circumstellar RHZs, the existence of circumbinary RHZs is not guaranteed. RHZs are based on inscribing spherical shells into the IHZ. Not all configurations permit this, however. If the inner RHZ borders intersect with the outer IHZ borders, RHZs vanish.

6.4 Orbital Stability

Orbital stability is a key requirement of potentially habitable configurations. The issue of the combined influence of orbital stability and insolation has been tackled by Huang (1960) in a quantitative manner. He used results of a toy model that is very popular in celestial mechanics, the circular restricted three-body problem, to constrain his HZ to dynamically "tame" regions. Since then, much work has been dedicated to investigating the stability of planets in binary star systems (e.g., Dvorak, 1986; Rabl and Dvorak, 1988; Whitmire *et al.*, 1998; Holman and Wiegert, 1999; Mardling and Aarseth, 2001; Pilat-Lohinger and Dvorak, 2002; Pilat-Lohinger *et al.*, 2003; Pichardo *et al.*, 2005; Doolin and Blundell, 2011; Jaime *et al.*, 2012; Georgakarakos, 2013). Among many other important results, it has been found that stable orbits are possible in the vicinity of circumstellar HZs (CSHZs) and circumbinary HZs (CBHZs).

In order to have an approximate idea on where one expects stable and unstable systems, one can resort to numerically generated fit functions (Dvorak, 1986; Rabl and Dvorak, 1988; Holman and Wiegert, 1999; Mardling and Aarseth, 2001). The fit functions of Holman and Wiegert (1999) for circumstellar and circumbinary motion are discussed in Chapter 2 in detail. In both cases a critical semi-major axis (a_{crit}) defines the border of stable planetary motion. Equation (2.2) tells us that the stable area around individual stars decreases for higher masses, higher eccentricities and smaller semi-major axes of the binary.

Comparing the black and the purple lines in Figure 6.1, one can see that the IHZ of the second star is only mildly truncated due to orbital instability. In contrast, the potentially habitable region around the primary is severely diminished by the presence of the second star. For two Sun-like stars on circular orbits, the IHZ borders can expand up to ≈25% of the single star HZ before orbital instability starts to chew away at the outer IHZ limit. No stable circumstellar orbits are left in the IHZ when the Sun-like binary stars orbit each other at a distance closer than 4 au. For double stars on highly eccentric orbits, the IHZ is substantially truncated even at large pericenter distances, minimizing the potential benefits of a second star regarding a system's habitability. Orbital stability constraints dictate that the maximum radiative contribution of the second star to the extent and displacement of the IHZ is relatively minor.

If the binary star orbit is tight enough, circumbinary IHZs are dynamically stable. Analytic estimates as to which configurations allow for dynamically stable HZs are discussed in Eggl (2018). As we see in the next section, however, dynamical stability and insolation geometry are not the only factors that determine where Earth-like planets can be habitable in a double star system.

6.5 Insolation Variability due to Orbital Dynamics

The fact that a double star interacts gravitationally with a planet in the same system has additional consequences for habitability. Due to gravitational perturbations caused by the second star, the planet's orbit evolves with time.

Hence, the planet-to-star distances change constantly. This has a profound impact on the amount of radiation a planet receives. In order to understand how much light arrives at a planet at any given time in a double star system, we introduce the momentary insolation function $\mathbb{I}(t)$. Neglecting variations in stellar luminosities ($dL_{A,B}/dt = 0$), the momentary insolation function $\mathbb{I}(t)$ is defined as the total spectrally weighted radiation arriving at the planet's upper atmosphere at time t:

$$\mathbb{I}(t) = \frac{A}{r_A^2(t)} + \frac{B}{r_B^2(t)}. \tag{6.8}$$

Equation (6.8) is a time dependent version of Equation (6.7). In Section 6.3.2, we looked at Equation (6.8) from a purely geometrical point of view, finding the distances r_A and r_B to Star A and Star B that do not exceed habitable insolation limits. Those distances determined the IHZ borders. Now we link r_A and r_B to the orbits of both the planet and the stellar binary.

In systems that are co-planar, we can express the star-to-planet distances in terms of Keplerian orbital elements as follows:

$$\mathbb{I}(t) = \frac{A}{r_P^2} + \frac{B}{r_P^2 + r_B^2 - 2r_P r_B \cos \Psi},$$

where

$$r_P = \frac{a_P(1 - e_P^2)}{1 + e_P \cos f_P}, \quad r_B = \frac{a_B(1 - e_B^2)}{1 + e_B \cos f_B},$$

$$\Psi = \varpi_P - \varpi_B + f_P - f_B. \tag{6.9}$$

Here, a_P and a_B are the planet's and the binary's semi-major axes measured in astronomical units, e_P and e_B are the corresponding eccentricities, and $f_P(t)$ and $f_B(t)$ are the planet and binary true anomalies, respectively. The planetary and the binary's longitudes of pericenter are denoted by ϖ_P and ϖ_B. We neglect the planet's perturbative effect on the orbit of the binary, since we are most interested in Earth-analogs. In hierarchical systems such as S-type and P-type binary-planet configurations, the planet's semi-major axis stays practically constant with time, while the planet's eccentricity vector does not (e.g., Marchal, 1990; Georgakarakos, 2003; Moriwaki and Nakagawa, 2004). In other words

$$\frac{da_P}{dt} = \frac{da_B}{dt} = \frac{de_B}{dt} = \frac{d\varpi_B}{dt} = 0.$$

In contrast, e_P and ϖ_P, as well as the true anomalies f_P and f_B, remain functions of time. Fortunately, explicit analytical expressions exist for all those quantities. The true anomalies, for instance, can be transformed into functions of time via Kepler's equation.

Simple secular evolution equations for the orbit of a massless particle in an S-type binary star system can be found in Heppenheimer (1978) and Andrade-Ines and Eggl (2017). Adopting a simple secular model, the orbital eccentricity evolution of a terrestrial planet in a binary star system is given by

$$e_P^2(t) = \eta^2 + \epsilon^2 + 2\eta\epsilon \cos(g\,t + \phi), \tag{6.10}$$

where η is the "free eccentricity", a component of the planet's eccentricity determined by the planet's initial orbit, and ϵ is the "forced-eccentricity", a component that is determined by the system parameters. Furthermore, g is the secular frequency dictating the timescale of the eccentricity evolution and ϕ is an initial phase. The forced-eccentricity and secular frequency and

$$\epsilon = \frac{5}{4}\frac{a_P}{a_B}\frac{e_B}{1 - e_B^2}, \qquad g = \frac{3}{4}\frac{m_B}{m_A}\left(\frac{a_P}{a_B}\right)^3 \frac{n_P}{(1 - e_B^2)^{3/2}}, \tag{6.11}$$

where $n_P = \sqrt{\mathcal{G}m_A/a_P^3}$ is the mean motion of the planet, $\mathcal{G}$ is the gravitational constant, and m_A is the primary star's mass. The evolution equation for the planet's longitude of pericenter is

$$\varpi_P(t) = \arctan\left(\frac{\eta \sin(g\,t + \phi)}{\eta \cos(g\,t + \phi) + \epsilon}\right). \tag{6.12}$$

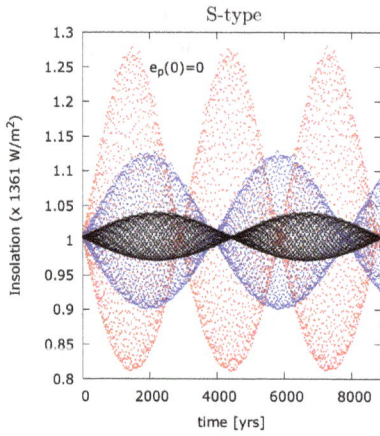

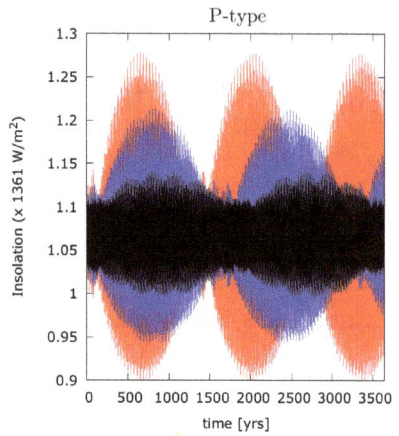

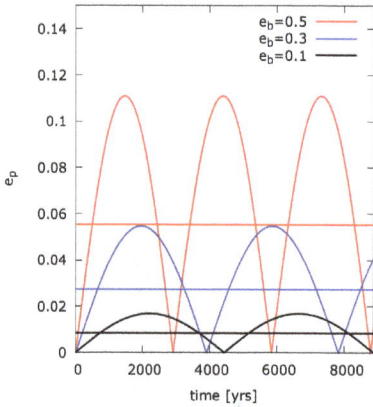

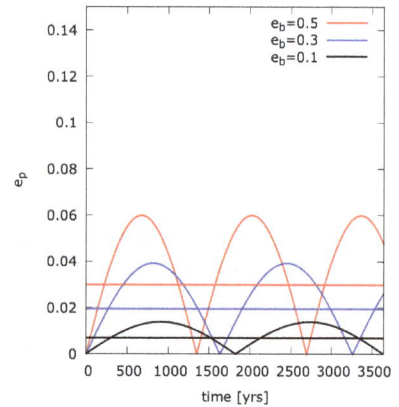

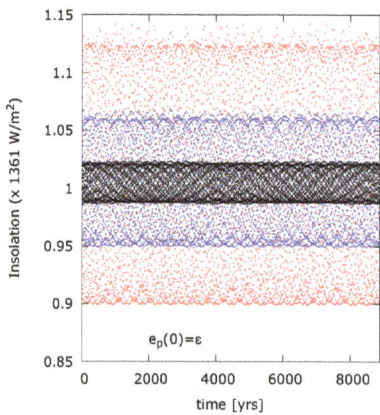

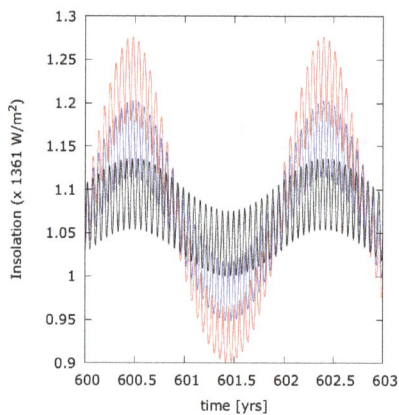

Planets that start out on a circular orbit have $e_P(0) = 0$. We must then require that $\phi = \pi$ and $\eta = \epsilon$. With the above assumptions, Equation (6.10) becomes

$$e_P(t) = \sqrt{2}\epsilon \, (1 - \cos(g \, t))^{1/2} . \qquad (6.13)$$

Equation (6.13) shows that even if a planet were to be on a circular orbit around its host star initially, its orbit would evolve into an ellipse due to the gravitational interaction with the second star. Figure 6.3 shows the evolution of the planet's orbital eccentricity as a function of time and the binary stars' orbital eccentricity. The changes in the planet's orbit drastically influence the amount of light the planet receives (Eggl *et al.*, 2012; Forgan *et al.*, 2015). The amplitude of the planet's eccentricity oscillation is determined by η and ϵ, as shown in Equation (6.10).

Perhaps somewhat counterintuitively, a planet on an initially circular orbit does not experience the smallest possible insolation variation. This comes from the fact that the eccentricity oscillates about the forced eccentricity, as can be seen from Equation (6.13). Thus, the dynamically least excited state corresponds to the planet's initial eccentricity being equal to the forced eccentricity,

$$\eta = 0; \qquad e_P(0) = e_P(t) = \epsilon.$$

In that case, the planet always retains the same eccentricity, namely the one that is injected by the second star. If the orbital eccentricity is equal

---

Figure 6.3: Momentary insolation ($\mathbb{I}$) and orbital eccentricity (e_P) evolution for an Earth-like planet in a circumstellar orbit ($a_P = 1$ au) around one component of a twin-Sun binary star system (G2V-G2V) with a semi-major axis of $a_B = 20$ au (left column) and in a circumbinary orbit ($a_P = 1.9$ au) around an F5V-G2V binary star system with a semi-major axis of $a_B = 0.4$ au (right column). Results for three eccentricities of the double star orbit are shown: $e_B = 0.1$ (black), 0.3 (blue) and 0.5 (red). The top panels present the insolation evolution for planetary orbits that were initially circular ($e_P(0) = 0$). The middle panels contain the corresponding evolution of the planet's orbital eccentricity (arcs). Additionally, the forced eccentricities (ϵ) in these systems are shown as horizontal lines. The insolation evolution for S-type systems where the planet is on a forced orbit ($e_P = \epsilon$) is shown in the lower left panel. The lower right panel contains a zoom of the top right panel, showing the rapid changes in insolation a circumbinary planet experiences. No spectral weights have been applied.

to the forced eccentricity, insolation variability, although still present, is less pronounced. This is shown in the bottom left panel of Figure 6.3. Planet formation theory suggests that planets in perturbed environments are more likely to form with eccentricities close to the forced eccentricity (e.g., Mardling, 2007; Silsbee and Rafikov, 2015). Orbital eccentricities of the majority of planets in binary star systems discovered so far are not always identical to the corresponding forced eccentricities (Bazsó *et al.*, 2017). S-type or P-type planets are, therefore, not necessarily in the most relaxed dynamical state.

For circumbinary planets, the momentary insolation function is

$$\mathbb{I}(t) = \frac{\mathbb{A}}{r_{bA}^2 + r_P^2 + 2r_{bA}r_P \cos \Psi} + \frac{\mathbb{B}}{r_{bB}^2 + r_P^2 - 2r_{bB}r_P \cos \Psi},$$

where $r_{bA} = \mu r_B$, $r_{bB} = (1-\mu)r_B$ and $\mu = m_B/(m_A+m_B)$. The distances r_B and r_P, and the mutual angle Ψ, can be calculated from Equations (6.9). Eccentricity evolution estimates for P-type orbits have been derived for instance by Moriwaki and Nakagawa (2004) and Georgakarakos and Eggl (2015). The forced eccentricity and secular frequency for the simplest model are

$$\epsilon = \frac{5}{4}\frac{a_B}{a_P}(1 - 2\mu)\frac{4e_B + 3e_B^3}{4 + 6e_B^2},$$

$$g = \frac{3}{4}\frac{n_B^2}{n_P}\left(\frac{a_B}{a_P}\right)^5 \mu(1 - \mu)\left(1 + \frac{3e_B^2}{2}\right).$$

Those can be inserted into Equations (6.12) and (6.13) to predict the orbit evolution of a single terrestrial circumbinary planet. Unlike in an S-type system, where the point of reference is the primary, a circumbinary orbit has the center of mass of the double star as the reference point.

Comparing the eccentricity evolution of circumbinary and circumstellar planets in Figure 6.3, we see that even after some time the investigated orbits differ only slightly from circular orbits. The consequences of the evolving orbits for a planet's insolation are drastic, however, exhibiting variation amplitudes of up to 50% of the initial value. Even though the planetary eccentricities in the circumbinary systems tend to be smaller than those in S-type systems, the amplitudes in insolation-variation remain comparable. That is because the combined radiation of both stars results in a steeper gradient of the insolation field for circumbinary planets. As a consequence, even small changes in the planet's orbit may have a large effect on the amount of light the planet receives. Note that in S-type systems, the large changes in insolation are not due to the radiation of the second star. Even during its pericenter passage, the second star barely contributes more than 1% to the total insolation. In fact, the gravitational interaction and the consequent changes in the planet's orbit are the main cause for the variable insolation conditions (Eggl *et al.*, 2012).

6.6 Dynamically Informed Habitable Zones

Planets in binary star systems can neither follow isophotes nor retain simple circular orbits over anything but the briefest periods of time. This entails that the amount of light a planet receives from the double star tends to change drastically with time.

How do planetary atmospheres react to potentially large variations in insolation? Simulating the climate of an Earth-like planet, Williams and Pollard (2002) found that oceans can buffer insolation variations effectively. They argued that the mean insolation received during one orbit is the only relevant quantity determining whether or not a planet is habitable. More recent studies indicate that the role of planetary eccentricity and the consequent insolation forcing cannot be discarded so easily. For instance, the long time spent in apocenter positions might cause a planet to fall into cold traps (e.g., Spiegel *et al.*, 2010; Dressing *et al.*, 2010). Forgan (2016) found that planets in binary star systems experience Milankovic cycles and that the orbital motion of the binary and the planet leave traces in the planet's mean surface temperature. Insolation variations, thus, are not all buffered. In particular, cold climatic states near the outer HZ border react strongly to variations in insolation, since the lack of greenhouse gases in the atmosphere and large ice-to-water fractions reduce thermal inertia (Popp and Eggl, 2017).

How quickly a planet's climate adapts to changes in insolation can be described via its "climate inertia". On planets with low climate inertia, surface temperatures follow changes in insolation quasi instantaneously, whereas a high climate inertia indicates that a planet's climate can buffer insolation variations to a certain degree. In order to study the effects of climate inertia on planetary habitability, Eggl *et al.* (2012) introduced three distinct types of HZs. If a planet's climate reacts quasi instantaneously to changes in insolation, the planet has to be inside the permanently habitable zone (PHZ) to allow for liquid water near its surface. For a planet to be in the PHZ, the maximum and minimum values of the insolation function must not exceed the habitable limits at any time. In this sense, the PHZ corresponds to the "classical" definition of a HZ. A more relaxed definition allows some parts of the planetary orbit to lie outside the PHZ. This defines the extended habitable zone (EHZ). However, most of the orbit must still fall within the PHZ. Following the argument of Williams and Pollard (2002) that the atmosphere buffers any variance in insolation, we can define the averaged habitable zone (AHZ). Insolation extrema are ignored as long as the time-averaged insolation stays within habitable bounds. Formally the three above HZs are defined as follows:

$$\text{PHZ: } \max_t(\mathbb{I}_I) \leq 1 \quad \text{and} \quad \min_t(\mathbb{I}_O) \geq 1,$$

$$\text{EHZ: } \langle \mathbb{I}_I \rangle_t + \sigma_I \leq 1 \quad \text{and} \quad \langle \mathbb{I}_O \rangle_t - \sigma_O \geq 1,$$

$$\text{AHZ: } \langle \mathbb{I}_I \rangle_t \leq 1 \quad \text{and} \quad \langle \mathbb{I}_O \rangle_t \geq 1.$$

Here, $\langle \mathbb{I} \rangle_t$ denotes the time-averaged momentary insolation function and σ^2 its variance. The EHZ always falls between the PHZ and AHZ borders. Here, we focus on deriving PHZ and AHZ borders only. We refer the reader to Eggl *et al.* (2012) for more details on calculating EHZs. Please note that all dynamically informed HZs do not depend on angular variables. Consequently, PHZ, EHZ and AHZ form concentric rings around the center of reference, just like the classical HZ, see Figure 6.4.

6.6.1   *Circumstellar habitable zones*

The most conservative HZ estimates for planets with unknown or very low climate inertia are provided via the PHZ. We can find PHZ limits for S-type systems by searching for configurations where the planet receives its insolation maximum and minimum, respectively. For the inner and outer

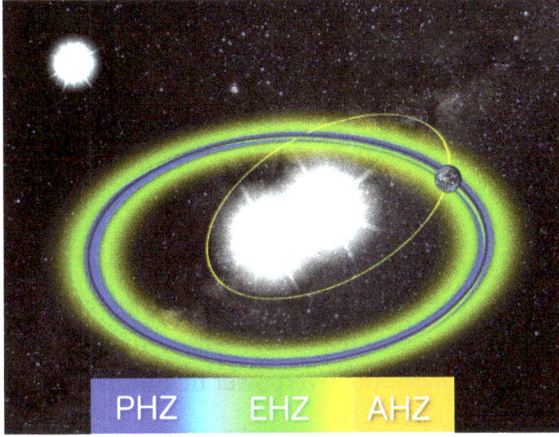

Figure 6.4: Dynamically informed HZs. The permanently habitable zone (PHZ, blue) corresponds to the most conservative estimate. A planet in the PHZ will never exceed habitable insolation limits. In contrast, a planet within the EHZ and AHZ experiences brief and ample excursions beyond SSHZ insolation limits. The average amount of light received over a planetary orbital period, however, must remain compatible with SSHZ insolation limits.

PHZ limits the following conditions hold:

$$\text{PHZ(I):} \quad \frac{\mathbb{A}_I}{q_P^2} + \frac{\mathbb{B}_I}{(q_P - q_B)^2} \leq 1,$$

$$\text{PHZ(O):} \quad \frac{\mathbb{A}_O}{Q_P^2} + \frac{\mathbb{B}_O}{(Q_P - Q_B)^2} \geq 1,$$

(6.14)

where $q_P = a_P(1 - e_P^{\max})$, $Q_P = a_P(1 + e_P^{\max})$, $q_B = a_B(1 - e_B)$ and $Q_B = a_B(1 + e_B)$. The maximum eccentricity the planetary orbit attains during its evolution is called $e_P^{\max} = \max_t [e_P(t)]$. The latter can be a non-trivial function of the binary's and the planet's orbital elements, as we shall see towards the end of this section. It is, therefore, often convenient to solve Equations (6.14) numerically with respect to a_P. The two solutions then correspond to the inner and outer PHZ limits. If the planet's initial orbit has a semi-major axis $\text{PHZ}(I) \leq a_P \leq \text{PHZ}(O)$, its orbit evolution will never carry it beyond the habitable insolation limits.

In systems where the binary does not influence the planet's orbit substantially, i.e., $e_P^{\max} \ll 1$, we can estimate the PHZ borders analytically via

$$\mathrm{PHZ}(I) \approx \frac{\mathbb{A}_I}{q_P(1 - e_P^{\max})} + \frac{a_P \mathbb{B}_I}{(q_P - q_B)^2}, \qquad (6.15)$$

$$\mathrm{PHZ}(O) \approx \frac{\mathbb{A}_O}{Q_P(1 + e_P^{\max})} + \frac{a_P \mathbb{B}_O}{(Q_P - Q_B)^2}, \qquad (6.16)$$

where we use $a_P(I, O) = (\mathbb{A}_{I,O})^{1/2}$ as an initial guess for the HZ borders to evaluate q_P, Q_P and $e_P^{\max}$. When formulating the maximum insolation condition for the inner edge of the PHZ, we have assumed that the radiative contribution of Star B does not overpower that of Star A. This condition formally reads

$$\mathbb{B}_I < (q_B - \sqrt{\mathbb{A}_I})^2.$$

If Star B is more luminous than dictated by this relation, the planet will most likely receive more light at its apocenter than at its pericenter and Equation (6.15) must be adapted accordingly

$$\mathrm{PHZ}(I) \approx \frac{\mathbb{A}_I}{Q_P(1 + e_P^{\max})} + \frac{a_P \mathbb{B}_I}{(Q_P - q_B)^2}. \qquad (6.17)$$

Next, we discuss how to determine HZ borders when the planet's climate has a high capacity to buffer changes in insolation. In that case, we calculate AHZ limits from insolation averages. In order to simplify the calculation of planetary insolation averages, Eggl *et al.* (2012) introduced the equivalent radii. Equivalent radii are constant distances with respect to the host star that yield the same average amount of insolation a planet would receive, were it on an elliptic orbit. In contrast to Eggl *et al.* (2012), here the equivalent radii $\bar{r}_P$ and $\bar{r}_B$ are chosen so as to be consistent with two body insolation averages. In other words,

$$\langle \mathbb{I}_A \rangle = \frac{1}{P} \int_0^P \frac{\mathbb{A}}{r_P^2(t)} dt = \frac{\mathbb{A}}{\bar{r}_P^2},$$

where $r_P(t) = a_P(1 - e_P^2)/(1 + e_P \cos f_P(t))$ and $\bar{r}_P = a_P(1 - \langle e_P^2 \rangle)^{1/4}$. The equivalent radius for the secondary star with respect to the host star is $\bar{r}_B = a_B(1 - e_B^2)^{1/4}$. Averaging over all possible planet binary configurations, we find that the conditions for the inner and outer border of the AHZ are as follows:

$$\text{AHZ}(I): \quad \frac{\mathbb{A}_I}{\bar{r}_P^2} + \frac{\mathbb{B}_I}{\bar{r}_B^2 - \bar{r}_P^2} \leq 1,$$

$$\text{AHZ}(O): \quad \frac{\mathbb{A}_O}{\bar{r}_P^2} + \frac{\mathbb{B}_O}{\bar{r}_B^2 - \bar{r}_P^2} \geq 1. \tag{6.18}$$

Note, that the above equations can be solved numerically for a_P to find precise values for the AHZ borders. As the average squared eccentricity, $\langle e_P^2 \rangle$, is very small in most cases, the following analytic estimate provides a good approximation:

$$\text{AHZ}(I, O) \approx \frac{\mathbb{A}(I, O)}{\bar{r}_P (1 - \langle e^2 \rangle)^{1/4}} + \frac{a_P \mathbb{B}(I, O)}{\bar{r}_B^2 - \bar{r}_P^2},$$

where $a_P(I, O) = (\mathbb{A}_{I,O})^{1/2}$ is our initial guess for the HZ, which is also used for estimating r_P and $\langle e^2 \rangle$.

Secular orbit evolution theory yields relatively compact expressions for $e_P^{\max}$ and $\langle e_P^2 \rangle$; see, for instance, Andrade-Ines and Eggl (2017). For planets on initially circular orbits, those expressions are as follows:

$$e_P^{\max} = 2\epsilon, \qquad \langle e_P^2 \rangle = 2\epsilon^2, \tag{6.19}$$

where ϵ is given by Equation (6.11). For dynamically less excited states, i.e., $e_P(0) = \epsilon$, we have

$$e_P^{\max} = \epsilon, \qquad \langle e_P^2 \rangle = \epsilon^2. \tag{6.20}$$

Note that eccentricity estimates based on first-order secular orbit evolution theory lack short period and resonant terms. More accurate estimates of the maximum and average squared eccentricity are available in literature (e.g., Eggl *et al.*, 2012; Georgakarakos, 2003; Andrade-Ines *et al.*, 2016).

6.6.2 Circumbinary habitable zones

Similar to S-type systems, the PHZ for circumbinary planets is calculated from those planet-star configurations that yield insolation extrema. The maximum insolation configuration occurs when the planet comes closest to the brightest star. The minimum insolation configuration has the brightest star farthest from the planet. Hence, we find

$$
\text{PHZ}(I): \quad \frac{\mathbb{A}_I}{(q_P - \mu Q_B)^2} + \frac{\mathbb{B}_I}{(q_P + (1 - \mu)Q_B)^2} \leq 1,
$$

$$
\text{PHZ}(O): \quad \frac{\mathbb{A}_O}{(Q_P + \mu Q_B)^2} + \frac{\mathbb{B}_O}{(Q_P - (1 - \mu)Q_B)^2} \geq 1,
$$

(6.21)

where $\mu = m_B/(m_A + m_B)$, Q_B is the distance between the two stars at apocenter and $\mathbb{A}_I > \mathbb{B}_I$. Here, as well as in S-type systems, the pericenter (q_P) and apocenter (Q_P) distances of the planet evolve with time. Insolation extrema are always related to maxima in the planet's orbital eccentricity with respect to time. In order to explicitly calculate the PHZ borders, Equations (6.21) can be solved numerically for a_P. The corresponding analytic estimates for small planetary orbital eccentricities are

$$
\text{PHZ}(I) \approx \frac{\mathbb{A}_I a_P}{(q_P - \mu Q_B)^2} + \frac{\mathbb{B}_I a_P}{(q_P + (1 - \mu)Q_B)^2},
$$

$$
\text{PHZ}(O) \approx \frac{\mathbb{A}_O a_P}{(Q_P + \mu Q_B)^2} + \frac{\mathbb{B}_O a_P}{(Q_P - (1 - \mu)Q_B)^2},
$$

(6.22)

with $a_P = r_{AB} = \sqrt{\mathbb{A}_{I,O} + \mathbb{B}_{I,O}}$ to be used as initial guess. In order to find circumbinary AHZ borders, we make use of equivalent radii. Defining

$$
\bar{r}_P = a_P(1 - \langle e_P^2 \rangle)^{1/4},
$$

$$
\bar{r}_{bA} = \mu a_B(1 - e_B^2)^{1/4},
$$

$$
\bar{r}_{bB} = (1 - \mu)a_B(1 - e_B^2)^{1/4},
$$

we find

$$
\langle \mathbb{I} \rangle_t \approx \frac{\mathbb{A}}{\bar{r}_P^2 - \bar{r}_{bA}^2} + \frac{\mathbb{B}}{\bar{r}_P^2 - \bar{r}_{bB}^2}.
$$

The equations for AHZ borders in circumbinary star systems then read

$$
\text{AHZ}(I, O) \approx \sqrt{\mathbb{A}_{I,O} + \mathbb{B}_{I,O}} \left(\frac{\mathbb{A}_{I,O}}{\bar{r}_P^2 - \bar{r}_{bA}^2} + \frac{\mathbb{B}_{I,O}}{\bar{r}_P^2 - \bar{r}_{bB}^2} \right).
$$

While the above formulae for PHZ and AHZ borders differ between S-type and P-type systems, Equations (6.19) and (6.20) still hold for $e_P^{\max}$ and $\langle e_P^2 \rangle$. The forced eccentricity, ϵ, also needs to be adapted to P-type systems. It is (Moriwaki and Nakagawa, 2004)

$$\epsilon = \frac{5}{4} \frac{a_B}{a_P} (1 - 2\mu) \frac{4e_B + 3e_B^3}{4 + 6e_B^2}.$$

6.6.3 *Application to α Centauri and Kepler-35*

We have explored several approaches to defining HZs in binary star systems, such as the IHZ, the RHZ and dynamically informed HZs. Is there a significant difference between those HZs? In order to answer this question, we shall make use of habitability maps.

Figures 6.5 and 6.6 contain habitability and maximum eccentricity maps for systems akin to α Centauri (S-type) and Kepler-35 (P-type). Habitability maps are excellent tools to investigate how different HZs behave in various double star systems. The two figures show how the PHZ, AHZ and RHZ borders change with increasing orbital eccentricity of the binary star. HZs, and uninhabitable and unstable regions, are color-coded in the left column. The RHZ and SSHZ borders are represented through vertical black and grey lines, respectively. A horizontal line shows the position of the actual α Centauri system (Figure 6.5), as well as the Kepler-35 system (Figure 6.6) in the map. For S-type A systems with low eccentricity, all HZs converge to the SSHZ. This is shown in the upper left panel of Figure 6.5. Not all binary stars move on circular orbits, however. If the double star has higher orbital eccentricity, the dynamically informed HZs start to diverge. With growing eccentricity of the double star orbit, the planet's orbit also becomes more eccentric and the PHZ starts to shrink. The actual α Centauri system, denoted by the horizontal grey line, has less than half of the planetary orbits in the SSHZ around α Centauri A permanently habitable. Assuming the planet is in its most relaxed dynamical state ($e_P = \epsilon$), the PHZ is somewhat larger but still only around 70% of the SSHZ. Table 6.2, which contains the various HZ borders for the actual α Centauri system, confirms this notion. Moreover, orbital stability starts to become an issue for planets with semi-major axes beyond 1.9 au in α Centauri A. The presence of the companion star is, thus, detrimental to the habitability of

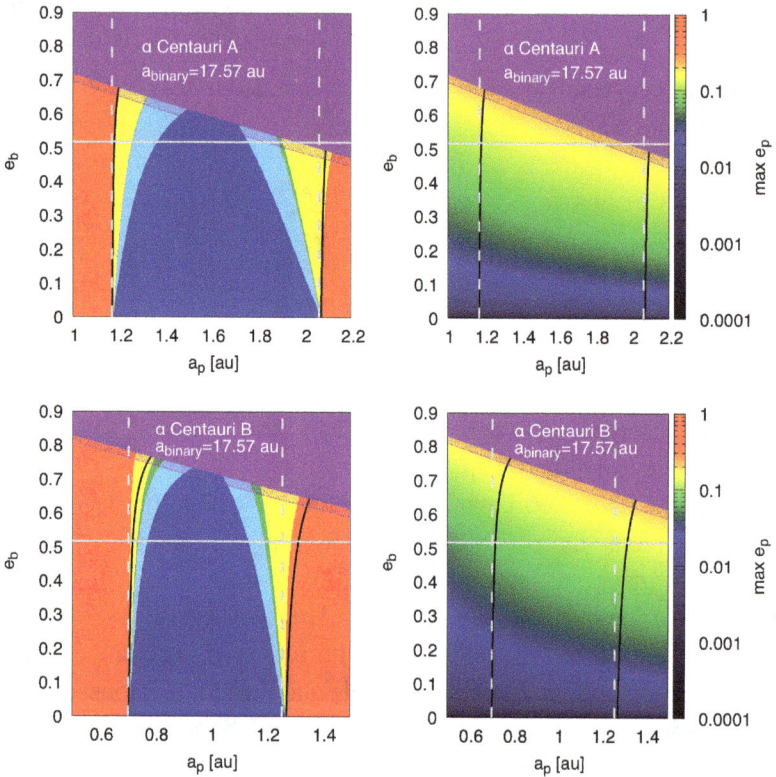

Figure 6.5: Habitability maps for α Centauri-like S-type systems (left column) and their corresponding maximum planetary eccentricities (right column) are shown as functions of the planet's semi-major axis (a_P) with respect to its host star and the binary star's orbital eccentricity. Red zones in the habitability maps are uninhabitable due to excessive or insufficient insolation. Yellow regions denote AHZs, EHZs are colored green, and dark blue zones represent configurations supporting permanent habitability (PHZs). Light blue zones are PHZs for the least excited dynamical configuration ($e_P = \epsilon$). Purple zones denote regions of orbital instability (Holman and Wiegert, 1999) (full), (Pilat-Lohinger and Dvorak, 2002) (dashed). The grey vertical lines denote SSHZ borders. The grey horizontal lines show the actual α Centauri system. The black full lines represent RHZ borders. Maximum eccentricity values are color-coded independently (see text for details).

potential planets around α Centauri A, especially if we assume the planet has low climate inertia. What if the planet's climate inertia was high instead? Comparing the system's AHZ to the SSHZ, we see that the presence of α Centauri B has very little effect on the HZ around α Centauri A. The constraints set by orbital stability remain the same no matter what climate

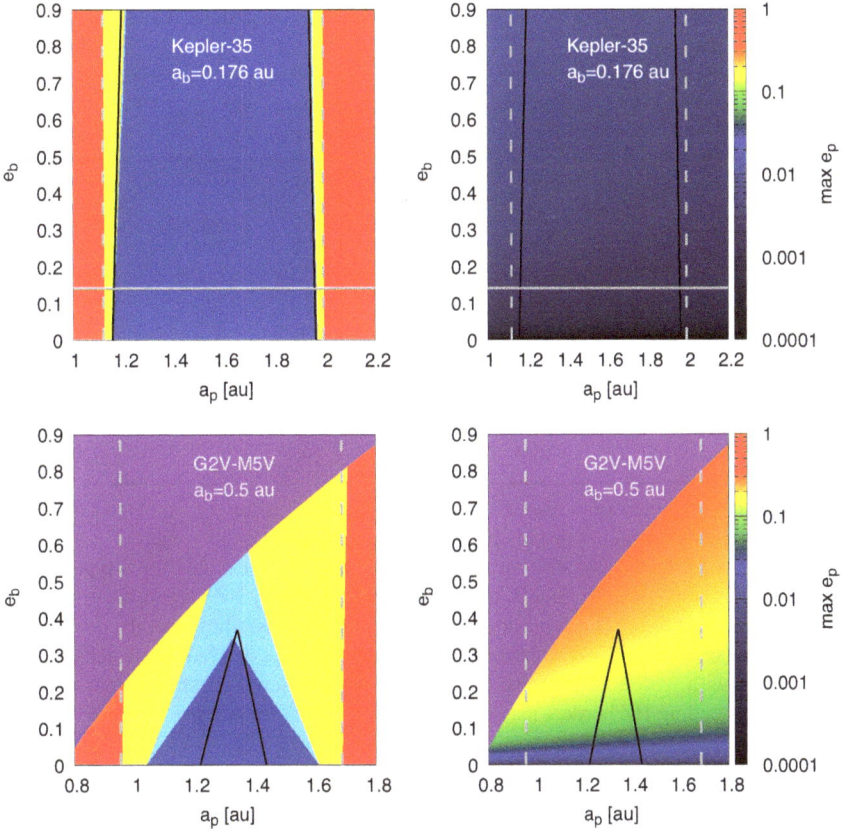

Figure 6.6: Same as Figure 6.5, only for P-type configurations. The upper panels show Kepler-35-like systems, whereas the lower panels show results for a G2V-M5V system with $a_B = 0.5$ au.

inertia the planet exhibits. For HZs around α Centauri B, similar features can be observed. Since the HZs are closer to the star, however, the onset of orbital instability is shifted to higher values of e_B. Some dynamically informed HZs around α Centauri B extend beyond the SSHZ, a consequence of Star A being substantially brighter than Star B. The corresponding RHZs were calculated assuming the stars are at the pericenter. If the RHZ calculations were performed with the stars close to their apocenter the RHZs limits would converge towards the SSHZ limits. For S-type systems such as α Centauri, the inner border of the RHZ falls between the AHZ and EHZ, whereas the outer RHZ border exhibits a behavior similar to the AHZ.

Table 6.2:   HZ borders for the α Centauri and Kepler-35 system. All HZ border values are given in [au].

System type	α Centauri S-type A	α Centauri S-type B	Kepler-35 P-type
$SSHZ_I$	1.171	0.706	1.121
$SSHZ_O$	2.064	1.265	1.994
RHZ_I	1.181	0.718	1.164
RHZ_O	2.092	1.315	1.958
PHZ_I	1.368	0.774	1.292
PHZ_O	1.760	1.136	1.956
PHZ_I^*	1.260	0.740	1.288
PHZ_O^*	1.892^i	1.194	1.960
AHZ_I	1.184	0.710	1.148
AHZ_O	2.130^i	1.290	2.009

Notes: In the case of Kepler-35, the SSHZ is derived using the combined flux of Kepler-35 A and B originating from the barycenter of the system. The HZ limits for S-type A systems are given with respect to Star A and for S-type B systems with respect to Star B, respectively. Circumbinary HZ borders are given with respect to the binary's barycenter. PHZ limits are assuming that the planet started on an initially circular orbit, whereas PHZ^* are derived for planets on orbits with forced eccentricity ($e_P = \epsilon$). The superscript i indicates that the corresponding HZ borders may be affected by orbital instability.

Circumbinary planets around Kepler-35-like double stars are less affected by orbital dynamics and climate inertia; this is shown in Figure 6.6. The PHZ and AHZ borders differ only marginally. Even high eccentricities of the binary star's orbit do not change HZ borders much. This is a consequence of having two Sun-like stars of similar mass. Such a configuration is known to suppress long term variations in a circumbinary planet's orbital eccentricity — a bonus for habitability. That this is not necessarily the case for all P-type systems is shown in the bottom panel of the same figure. Here, a circumbinary configuration is shown that favors eccentricity injection into the planet's orbit. The greater the disparity of the masses of the double star components and the higher the binary orbital eccentricity, the lower the chances for a circumbinary planet with low climate inertia to remain in a habitable state. For G2V-M5V systems with $a_B = 0.5$ au and $e_B > 0.35$ the PHZ vanishes completely, if the planet's orbit is initially circular. Only considerable climate buffering capabilities guarantee the habitability of circumbinary worlds in such systems. Figure 6.6 shows, moreover, that the circumbinary RHZ is closer to the circumbinary PHZ while the RHZ is more likely to align with the inner border of the EHZ and the outer border of the AHZ in S-type systems (Figure 6.5).

6.7 Self-consistent Models

We have seen that combining analytic insolation estimates with precomputed spectral weights allows for a quick assessment of where habitable worlds can be expected in binary star systems. This approach does have its limits, however.

Orbit evolution models based on the three-body problem, for instance, do not account for additional perturbers. Other planets in the system alter the orbit of the potentially habitable planet. More advanced dynamical models are required in order to include those effects (Forgan, 2016; Bazsó *et al.*, 2017). The influence of obliquity and spin state on a planet's habitability cannot be studied using the methods presented in this chapter.

In order to investigate such phenomena, self-consistent simulations of a planet's climate and orbital evolution become necessary. Coupling orbit propagators to longitudinally-averaged energy balance models (LEBMs) and general circulation models (GCMs), Forgan (2016) and Popp and Eggl (2017) have shown that the variable insolation a planet is exposed to in double star systems leaves traces in the global surface temperature, precipitation and glaciation. In how far this affects habitability remains difficult to judge.

Self-consistent HZ calculations are time-consuming and tuned to a specific climate model. Hence, results have to be interpreted with care (Forgan, 2014; Popp and Eggl, 2017). Nevertheless, Popp and Eggl (2017) argue that variations in insolation can be more easily buffered in warm states, close to the inner edge of the classical HZ. Cold climate states appear to be much more sensitive to variations in insolation. This suggests that the AHZ could be a reasonable proxy for inner border of HZs in binary star systems, whereas the PHZ or EHZ may be more suitable to describe outer HZ limits.

6.8   Super-habitability of Binary Star Systems

Our Solar System has produced a habitable world, but that is not to say that conditions for planets to be habitable are not better elsewhere. In fact, revisiting Figures 6.5 and 6.6 we see that certain S-type binary star systems may harbor larger HZs than equivalent single stars. Mason *et al.* (2013, 2015) found another argument for super-habitability of P-type systems. Investigating the stellar evolution of close binary stars they claim that certain stellar configurations suppress stellar activity. This, in turn, reduces the XUV flux and solar wind pressure which are ultimately responsible for the photolysis of water in the atmosphere and consequent loss of hydrogen to space (e.g., Kislyakova *et al.*, 2013; Lichtenegger *et al.*, 2016). Johnstone *et al.* (2015c) found that circumbinary planets have to cross shock fronts created by the stellar winds of the binary on a regular basis, however, which may have adverse effects on their habitability. Certainly, more research is needed to draw a clearer picture of whether or not binary star systems can be super-habitable.

6.9   Summary

Earth-like planets in binary star systems can be habitable. The presence of a second source of radiation, the motion of the double star itself and the gravitational interaction with the planet, however, continuously change the amount and spectral composition of light arriving at the planet. Classical single star HZ estimates do not account for such effects. As a consequence, SSHZ estimates can fail spectacularly in predicting where to look for habitable worlds in double star environments. Several types of modified HZs have been proposed to tackle this issue; those are summarized in Table A.1. Using dynamically informed HZs, we have shown that in systems akin to α Centauri, a planet's climate inertia, i.e., the reaction of a planet's climate to changes in insolation, plays a crucial role in determining where potentially habitable worlds are to be found. The habitability of circumbinary planets around two equally massive stars is much less affected by a planet's climate inertia, thus making such systems excellent candidates in the search for habitable worlds. This is not the case for all P-type systems, though. If the components of the binary differ in mass and luminosity, circumbinary planets experience pronounced changes in insolation on short timescales so that climate inertia becomes a critical factor for habitability again. Self-consistent calculations using LEBMs and GCMs suggest that the reaction of a planet's climate to changes in insolation may not be the same for the inner and outer edges of the HZ as climate inertia differs in cold and hot climates. Determining a planet's climate inertia is, thus, vital to better understand where habitable planets can be expected in binary star systems.

Chapter 7

Habitability of Known Planets in Binary Star Systems

This chapter presents a collection of binary star systems with detected extrasolar planets, either of circumstellar or circumbinary type. These systems allow us to combine our methods described in previous chapters, and to apply them to study the dynamics and the habitability of the planets.

7.1 Discovered Exoplanets in Binary Systems

In the past two decades, the number of detected extrasolar planets (or exoplanets for short) has increased to well more than 3,000 with about a similar number still waiting to be confirmed. There exist several catalogues of exoplanets that collect the data and make it publicly available for free. Among those catalogues, the two most frequently used are the Extrasolar Planet Encyclopaedia[1] (Schneider *et al.*, 2011) and the Exoplanet Orbit Database[2] (Han *et al.*, 2014). Another rather specialized catalogue is the Catalogue of Exoplanets in Binary Star Systems[3] (Schwarz *et al.*, 2016) that specifically lists only exoplanets in binary and multiple star systems, unlike the former two more general catalogues. Such a separate listing of exoplanets in binary (and multiple) star systems, together with their

---

[1] http://exoplanet.eu.
[2] http://exoplanets.org.
[3] http://www.univie.ac.at/adg/schwarz/multiple.html.

dynamical type (P-type or S-type, see Chapter 2), facilitates the study of those planets.

For the following sections, we restrict ourselves to exoplanets in binary star systems and give an overview of their physical and dynamical properties (if known). The data was taken from the catalogue of Schwarz *et al.* (2016), that itself has cross-references to other catalogues.

S-type systems

Table 7.1 collects 43 S-type exoplanets of 42 different binary star systems. This sample of S-type exoplanets does not include multi-planetary systems, which would make the list more extensive. The stellar separations scatter widely and range from several thousand au for the wide binaries to just a few au at the lower end. It is remarkable that the eccentricity of the binary (e_B) is known in very few cases, although this is one of the key parameters for the study of the system's dynamical stability. For future observations, one of the top priorities should be to determine the secondary star's orbital parameters; however this is a long term task due to the very long orbital periods of the wide binaries. The naming scheme used indicates the planet hosting star, e.g., "Ab" when the planet orbits Star A, or "Bb" when the planet orbits Star B.

P-type systems

Table 7.2 includes the complete list (as of December 2017) of all 23 discovered P-type exoplanets in 20 different binary star systems. They are ordered by decreasing separation, a_B (in au), between the stellar components A and B. Stellar masses, m_A and m_B, are in units of the solar mass ($M_\odot$), while the planet masses are given in units of Jupiter's mass ($\approx 10^{-3} M_\odot$). In general, the two stars are close binaries separated by 0.25 au or less, except for FW Tau where the stellar components are 11 au apart. The P-type planets are orbiting the stars at a distance of at least 3 times the stellar separation ($a_P > 3a_B$). As an observational fact, it gives a hint that planetary mass objects can only move in stable P-type motion beyond the 5:1 MMR with the inner binary. This hypothesis is strengthened by the numerical results of Dvorak (1986) and Holman and Wiegert (1999).

Table 7.1: List of S-type binary star systems ordered by decreasing stellar separation a_B. Stellar masses, m_A and m_B, have units of $M_\odot$, and the planet's mass, m_P, is given in units of Jupiter's mass; the distances a_B and a_P are in au. For details about individual systems, see Section 7.4.1.

Name	m_A	m_B	m_P	a_B	a_P	e_B	e_P
HIP 70849 Ab B	0.63	0.05	9	9,000	10	—	0.6
HD 222582 AB	0.99	0.3	7.75	4,746	1.35	—	0.725
HD 147513 Ab B	0.92	0.65	1.21	4,451	1.32	—	0.26
HD 213240 Ab C	1.22	0.146	4.5	3,898	2.03	—	0.45
HD 101930 Aa B	0.74	0.666	0.3	2,200	0.302	—	0.11
HAT-P-1b (ADS16402 A Bb)	1.12	1.16	0.524	1,550	0.0554	—	0.067
HD 80606b / HD 80607	0.98	0.98	3.94	1,200	0.921	0.5	0.934
WASP-70 Ab B	1.11	0.8	0.59	810	0.0485	—	0
HD 142022 Ab B	0.99	—	5.1	794	3.03	—	0.53
Kepler-432 Ab B	1.35	—	5.41	760	0.301	—	0.512
TrES-4 Ab C	1.18	0.59	0.917	750	0.0508	—	0
HD 188015 Ab B	1.09	—	1.26	684	1.19	—	0.15
HD 75289 Ab B	1.05	0.14	0.42	620	0.046	—	0.024
HD 109749 Ab B	1.1	0.78	0.28	390	0.0635	—	0.01
HD 46375 Ab B	0.91	—	0.249	314	0.041	—	0.04
WASP-77 Ab B	1.00	0.71	1.76	306	0.024	—	0
KELT-2 Ab B	1.317	0.78	1.486	295	0.055	—	0.185
HD 114729 Ab B	0.93	0.253	0.84	282	2.08	—	0.32
Kepler-14 Ab B	1.51	1.39	8.4	280	0.08	—	0.035
HD 27442 Ab B	1.2	0.6	1.35	240	1.16	—	0.058
TrES-2 Ab B	0.98	0.509	1.253	232	0.0356	—	0
HD 212301 Ab B	1.27	0.35	0.4	230	0.036	—	0.015
HD 16141 Ab B	1.01	0.286	0.215	220	0.35	—	0.28
HD 189733 Ab B	0.8	0.2	1.138	216	0.0314	—	0.004
Kepler-410 Ab B	1.21	—	—	211	0.1236	—	0.17
WASP-85 Ab B	1.04	0.88	0.969	205	0.038	—	0.103
GJ 15 Ab B	0.38	0.16	0.017	150	0.0717	—	0.12
HD 114762 Ab B	0.84	0.1383	10.98	132	0.353	—	0.335
HD 195019 Ab B	1.02	—	3.7	131	0.1388	—	0.014
WASP-2 Ab B	0.89	0.48	0.847	106	0.0314	—	0
HD 19994 Ab B	1.34	0.90	1.68	100	1.42	0.26	0.3
GJ 3021 Ab B	0.9	0.13	3.37	68	0.49	—	0.511
OGLE-2008-BLG-092L	0.58	0.12	0.137	54	18	—	—
τ Bootis Ab B	1.3	0.4	5.95	45	0.046	0.91	0.023
WASP-11 Ab B/HAT-P-10 b	0.82	0.34	0.46	42	0.0439	—	—
HD 126614 Ab B	1.15	0.32	0.36	36	2.35	0.6	0.41
HD 41004 Ab B	0.7	0.4	2.54	23	1.64	—	0.39
HD 41004 A Bb	0.7	0.4	18.4	23	0.0177	—	0.058
HD 196885 Ab B	1.33	0.45	2.98	23	2.6	0.42	0.48
GJ 86 Ab B	0.8	0.5	4.01	21	0.11	0.4	0.046
γ Cep B	1.4	0.4	1.85	20	2.05	0.41	0.049
OGLE-2013-BLG-0341 Bb	0.113	0.121	0.005	12	0.702	—	—
Kepler-420 Ab B	0.99	0.70	1.45	5.3	0.382	0.31	0.772

Table 7.2:   List of P-type binary star systems ordered by decreasing stellar separation a_B. Stellar masses, m_A and m_B, have units of $M_\odot$, and the planet's mass, m_P, is given in units of Jupiter's mass; the distances a_B and a_P are in au. For details about individual systems, see Section 7.4.2.

Name	m_A	m_B	m_P	a_B	a_P	e_B	e_P
FW Tau AB b	0.28	0.28	10	11	330	—	—
Kepler-34 AB b	1.048	1.021	0.22	0.2288	1.0896	0.521	0.182
Kepler-16 AB b	0.6897	0.2026	0.333	0.2243	0.7048	0.159	0.007
PSR B1620-26	1.35	0.6	2.5	0.2	23	0.025	—
Kepler-453 AB b	0.934	0.194	0.03	0.1848	0.788	0.051	—
Kepler-35 AB b	0.888	0.809	0.127	0.1762	0.6035	0.142	0.042
Kepler-38 AB b	0.949	0.249	0.38	0.147	0.4644	0.103	—
Kepler-1647 AB b	1.22	0.97	1.52	0.1276	2.72	0.1602	0.058
Kepler-413 AB b	0.82	0.5432	0.211	0.1015	0.353	—	0.118
Kepler-47 AB b	1.043	0.362	—	0.0836	0.2956	0.0234	—
Kepler-47 AB c	1.043	0.362	—	0.0836	1	0.0234	—
OGLE-2007-BLG-349 AB b	0.41	0.3	0.25	0.08	2.59	—	—
RR Cae AB b	0.44	0.182	4.2	0.0076	5.3	—	—
NY Vir AB b	0.46	0.14	2.78	0.0044	3.39	—	—
NY Vir AB c	0.46	0.14	4.49	0.0044	7.54	—	—
Kepler-451 AB b	0.48	0.12	1.9	0.0041	0.92	—	—
HU Aqr AB b	0.88	0.2	5.9	0.0039	6.18	0	0.29
NN Ser AB b	0.535	0.111	2.28	0.0039	3.39	—	0.2
NN Ser AB c	0.535	0.111	6.91	0.0039	5.38	—	0
DP Leo AB b	0.6	0.009	6.05	0.0027	8.19	0	0.39
SR 12 AB c	0.3	—	13	—	1083	—	—
Ross 458 AB c	0.6	0.075	8.5	—	1168	—	—
ROXs 42 AB b	0.89	0.36	10	—	140	—	—

7.2   Habitable Zones

In this section, we calculate the habitable zones (HZs) for some of the systems from Tables 7.1 and 7.2.

7.2.1   *HZ of S-type systems*

For S-type systems, we restrict ourselves to binaries with $a_B < 500$ au and ignore the wider separation cases. Desidera and Barbieri (2007) showed by a statistical analysis that the properties of exoplanets of wide binaries ($\gtrsim 300$ au) are compatible with those of planets orbiting single stars. They concluded that in wide separation binaries, the secondary star has only a very limited influence on planet formation and dynamical evolution. In contrast to that, Kaib *et al.* (2013) found that even for very wide binaries (with $a_B > 1,000$ au) outside perturbations of the binary's orbit by passing stars or the galactic tide can lead to the disruption of planetary systems (on time-scales of 10^8-10^9 years).

In any case, the companion star's influence on the total effective insolation onto the planet becomes negligible for all but the tightest binaries. Accordingly, here we calculate the HZ boundaries based on only the insolation from the host star and completely ignore the insolation from the secondary. We apply the model of Kopparapu *et al.* (2014) for main-sequence stars with effective temperatures in the range of 2600–7200 K and a planet of one Earth mass ($M_\oplus$). However, there are still some caveats concerning the stellar parameters:

(1) The model assumes the host star to be on the main sequence, which is not fulfilled, for instance, for γ Cep, HD 27442, and a few more. In such cases we calculate the effective stellar flux S_{eff} (which is required for the model) from the published effective temperature for the respective star, but with the increased luminosity for the post-main sequence phase. A remedy to handle these cases would be to use an adjusted model for post-main sequence stars, like in the work of Ramirez and Kaltenegger (2016).

(2) When the mass of the secondary star is unknown (e.g., HD 46375) we do not calculate the HZ at all.

(3) In some cases, the star's mass has been published, but not its radius (e.g., HD 212301 and OGLE systems). Here we make use of the mass-radius relation to approximate the star's luminosity, where we also take into account the constraint by the effective temperature.

The results are shown in Figure 7.1. It depicts the location and extent of the HZ (in blue) for 27 different stars whose names are given in the left margin. At the right margin, we include the spectral type of the host star (with the implicit assumption that it is a main sequence star). Black dots show the location of the exoplanet relative to the host star. The horizontal red error bars mark the region where a secular resonance (SR) with the planet can occur (see Section 7.3.2 for details).

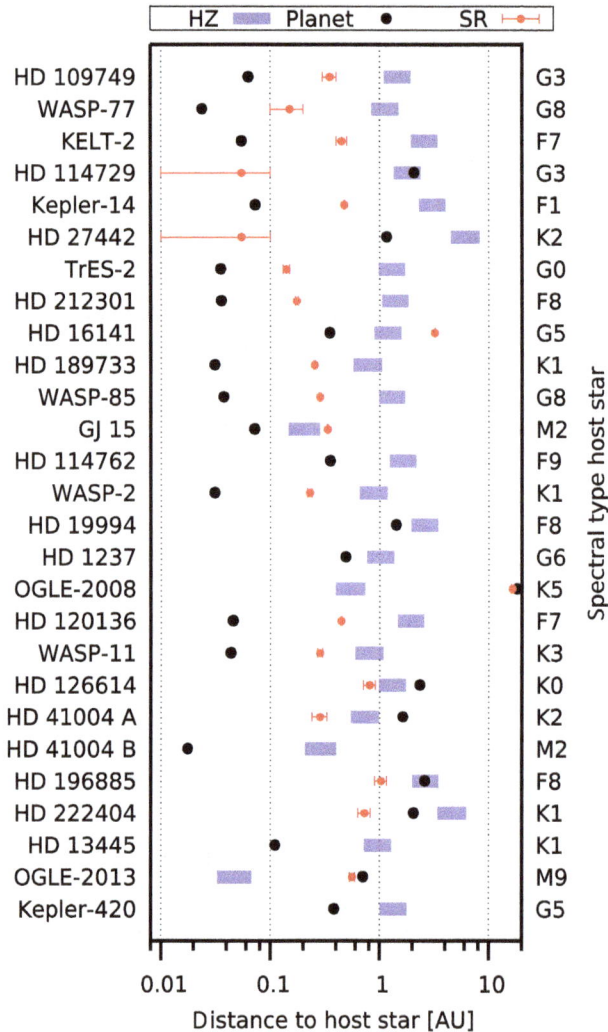

Figure 7.1: Habitable zones for S-type systems with stellar separations below 500 au. The blue boxes indicate the sizes of the HZs, while the black dots show the positions of the exoplanets. Errorbars mark the expected regions where SRs can occur.

From the figure, it is obvious that in this sample, only two out of 27 detected planets orbit in the HZs of their host stars (HD 114729 and HD 196885), while a few other are located close to it. About half of all planets move in close-in orbits with $a_P \leq 0.1$ au. This can be explained in part by observational bias, i.e., by detecting preferentially massive, close-in planets with large radii by radial-velocity and transit measurements (Benz *et al.*, 2008; Mills and Mazeh, 2017). In contrast, most detected planets seem to be comfortably far from the HZ to allow for additional undetected habitable planets.

Now we can ask whether there exists only the single discovered planet or if there could exist yet undiscovered planets. To check this hypothesis, we have to determine the locations of (linear) SR. If the SR were inside or close to the HZ, then it would prohibit other planets from being there. On the other hand, if there were no SR in the HZ, then that system could be a promising candidate for further observations.

Before going into the details of how to find an SR, we first need to check whether our assumption that the HZ is entirely determined by the insolation from the host star is correct. We have to compare the extent of the HZ for the host star alone (classical or single star HZ) relative to the extent when the radiation from both stars is considered (binary star HZ).

This can be achieved by calculating the averaged HZ (AHZ) from the model of Eggl *et al.* (2012) (cf. Section 6.6). We combine that model with the model of Kopparapu *et al.* (2014) to obtain the effective stellar flux, $S_{\text{eff},i}$, for both stars ($i = 1, 2$). The value $S_{\text{eff},i}$ depends on the luminosity and effective temperature of the respective star. In Figure 7.2, we can see that the agreement between the two estimates is excellent — the blue and green boxes match. At the right margin in the figure, the spectral type of the distant companion star is shown, again presumed to be a main sequence star.[4] There is virtually no difference in the extent of the HZ for all but the tightest binaries ($a_B < 20$ au). Table 7.3 compares the single star HZ (SSHZ) and the AHZ for binary systems with separations below 100 au. The numbers in the table are given with three digits of accuracy to highlight the minuscule differences between the two approaches. On comparing the inner and outer HZ borders it becomes evident that the outer borders are more susceptible to changes (as expected), as they expand when an additional source of radiation is present. This comparison demonstrates that for the

---

[4]Note that the companion of GJ 86 (HD 13445) is a white dwarf, which obviously violates the assumption in this specific case.

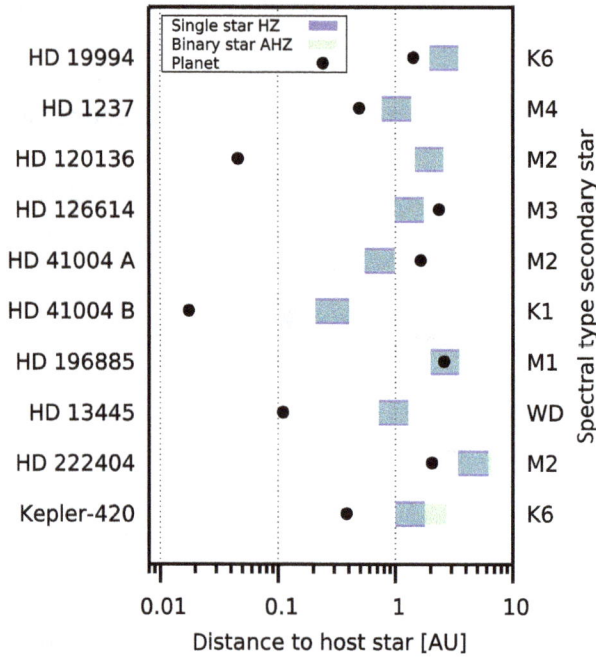

Figure 7.2: Comparison of classical HZ for single stars (in blue) with AHZ for binary stars (in green). Only binaries with separation less than 100 au are checked.

Table 7.3: HZs of selected S-type binary systems with $a_B \leq 100$ au. Single star HZs are based on the formula of Kopparapu *et al.* (2014), while binary star AHZ are based on Eggl *et al.* (2012); see text for details.

Name	Single star HZ	Binary star AHZ
94 Cet (HD 19994)	1.949–3.419	1.949–3.420
GJ 3021 (HD 1237)	0.765–1.359	0.765–1.358
τ Boo (HD 120136)	1.464–2.552	1.466–2.562
HD 126614	0.981–1.739	0.981–1.744
HD 41004 A	0.542–0.980	0.543–0.982
HD 41004 B	0.207–0.399	0.207–0.400
HD 196885	1.985–3.458	1.995–3.503
GJ 86 (HD 13445)	0.716–1.279	0.717–1.282
γ Cep (HD 222404)	3.395–6.195	3.446–6.554
Kepler-420	0.998–1.774	1.009–2.688

Table 7.4: HZs of P-Type systems. HZ were calculated with the formula of Kopparapu *et al.* (2014), unless stated otherwise. For the last seven systems, missing parameters do not allow us to calculate the HZ.

Name	HZ [au]
FW Tau AB	0.57–1.11
Kepler-34 AB	1.37–2.40
Kepler-35 AB	1.20–2.13
Kepler-1647 AB	1.42–2.49
Kepler-413 AB	0.87–1.59
NY Vir AB	0.24–0.47
OGLE-2007-BLG-349 AB[a]	0.64–1.26
Ross 458 AB (DT Vir AB)[a]	0.68–1.33
Kepler-16 AB[b]	0.74–1.37
Kepler-453 AB[b]	0.92–1.63
Kepler-38 AB[b]	0.92–1.63
Kepler-47 AB[b]	0.99–1.75
ROXs 42AB[b]	0.97–1.99
PSR B1620-26 / WD J1623-266	—
RR Cae AB	—
Kepler-451 AB / 2M 1938+4603	—
HU Aqr AB	—
NN Ser AB	—
DP Leo AB	—
SR 12 AB	—

[a]For these systems the HZs are calculated with the formula of Kaltenegger and Haghighipour (2013).
[b]HZs are calculated using the formula for single stars and the combined temperatures and luminosities of both stars.

still wider binaries the insolation of the second star becomes practically negligible.

7.2.2 *HZ of P-type systems*

In Table 7.4, we show the HZs of the P-type systems from Table 7.2. The first six systems were calculated with the formula of Kopparapu *et al.* (2014), while the next two systems (OGLE-2007 and Ross 458) were determined using the formula of Haghighipour and Kaltenegger (2013). Furthermore, for the next five entries we applied the formula for single stars with the combined temperature and luminosity of the two components. For the last seven binaries, there are not enough parameters known to calculate the HZ.

7.3　Gravitational Perturbations

We have discussed in Chapter 3 that resonant perturbations can have a severe effect on the stability of a planetary system. Both MMR and SR can occur for S-type planets. Let us for now focus only on the latter type of resonances.

7.3.1　*Secular resonances*

Returning to the selected S-type systems from Table 7.1, we can estimate the secular precession frequency of the known planet by using the methods from Section 3.3. Once that secular frequency is known, we determine the locations where massless test planets would have exactly the same precession frequency, and thus would suffer from a secular resonance.

This procedure seems simple enough, but some complications arise from the facts that (i) all masses must be known (both stars and planet), (ii) in most cases the secondary star's eccentricity is unknown, and (iii) sometimes even the planet's eccentricity is missing.

We narrow the sample by posing two restrictions. Firstly, the binary separation should be <500 au. This is for the same reason as explained above for the HZ; the wider binaries can be well approximated as single stars. Secondly, if one of the masses is unknown, then we discard that system from the analysis.

Regarding the secondary's eccentricity, we consistently use as limiting cases $e_B = 0$ (as a lower bound) and $e_B = 0.5$ (as an upper bound). Note that a circular orbit for the binary is unrealistic, like the few examples from Table 7.1 show for which e_B could be determined from observations. However, for the value $e_B = 0$ the secular period of the planet will be maximal and thus the SR will be located the farthest away from it. The other value ($e_B = 0.5$) is derived from the observed eccentricity distribution of binaries with periods >1,000 days (Duquennoy *et al.*, 1991; Raghavan *et al.*, 2010). In this case, the planet's secular period is smaller and the SR is closer to it.

In most of the cases we can use the nominal value for the planet's eccentricity (see Table 7.1). However, if e_P is unknown, then we use the four different values 0.001, 0.1, 0.2 and 0.5 to cover the most likely parameter range for the planet's eccentricity.

For every possible combination of e_B and e_P (the other parameters are held fixed), we apply the semi-analytical method and the formula of Georgakarakos (2003) to calculate the planet's secular frequency. In this way we obtain between two to eight estimates of the frequency, from which we determine its minimum and maximum values. These two values then define a semi-major axis interval that probably contains the SR with the detected planet (within the uncertainties due to the observational errors); this interval is represented by red error bars in Figure 7.1.

7.3.2 Additional effects

We have already mentioned that about half of all the planets in Table 7.1 have semi-major axes $a_P \le 0.1$ au. At such small distances to the host star, a planet's secular precession frequency, g_P, is small, since it scales as $g_P \propto \alpha^3$ with the semi-major axis ratio $\alpha = a_P/a_B$. So for large a_B, it naturally follows that $\alpha \ll 1$, and then g_P is too small (the secular period is too large) for any (linear) SR to occur.

So far we have dealt with classical Newtonian physics, but it is well known from Einstein's theory of General Relativity (GR) that in the vicinity of large masses, space-time is curved and additional perturbations act. One such effect is the relativistic precession of the perihelion (GRP), which was one of the first test cases for the theory. Einstein could explain the additional precession of Mercury's perihelion of 43 arc-seconds per century, which was impossible to derive before from Newton's theory.

It is possible to include the effects of general relativity by adding additional potential terms to the equations of motion and then solving them numerically. Instead, here we apply a simplified analytical formula taken from Brasser *et al.* (2009) that is able to mimic the effect of GR and yields the average variation in the longitude of the pericenter

$$g_{GR} = \left\langle \frac{d\varpi}{dt} \right\rangle_{GR} = 0.0383 \left(\frac{1 \text{ au}}{a_P} \right)^{5/2} \left(1 - e_P^2\right)^{-1}.$$

From this formula, we get the relativistic precession frequency, g_{GR}, in units of arc-seconds per year. For more details about the origin of the numerical factor and how to derive this expression from the perturbing potential, see Brasser *et al.* (2009).

Returning to Figure 7.1, we combine the classical and relativistic contributions to the secular frequency, $g_{sec} = g_P + g_{GR}$. The first term dominates for planets with roughly $a_P > 1$ au, while the second term dominates for those with $a_P < 0.1$ au; between 0.1 and 1 au, the two contributions are of comparable magnitude. Then we plot with red error bars the range between the minimum and maximum values of g_{sec} (corresponding to $e_B = 0$ and $e_B = 0.5$, respectively). In the region marked in this way, a terrestrial planet would be at risk of becoming trapped in a linear SR with the detected planet. As we can see from the figure, such an SR cannot occur in some systems, e.g., HD 114762, HD 19994 and HD 1237. In these cases, the secondary star is not too distant ($a_B < 150$ au) and that the planet is located interior to the HZ. For two systems (GJ 15 and HD 126614) the SR is next to the HZ and can cause highly eccentric motion.

7.3.3 *Discussion*

So far we have not mentioned another important source of perturbations, namely one or more additional massive planets. Since up to now we discussed only binary star systems with a single giant planet, we should consider what would happen with at least two giant planets. In the Solar System, Jupiter and Saturn are the key drivers of the secular dynamics by virtue of their masses. It is well known that the mutual gravitational interaction between them causes strong SR in the main-belt of asteroids (Froeschlé and Scholl, 1989), and eventually even in the terrestrial planet region (Pilat-Lohinger *et al.*, 2008a,b). Similar to the situation in the Solar System, the presence of another giant planet would increase the secular frequencies of both planets, which then would give rise to multiple locations of SR. Hence, in multi-planetary systems, secular resonances are a common phenomenon, no matter if for single or multiple stars.

Figure 7.3 sketches the two principal cases of the relative arrangement of a giant planet and the HZ. At the top we have the interior case where

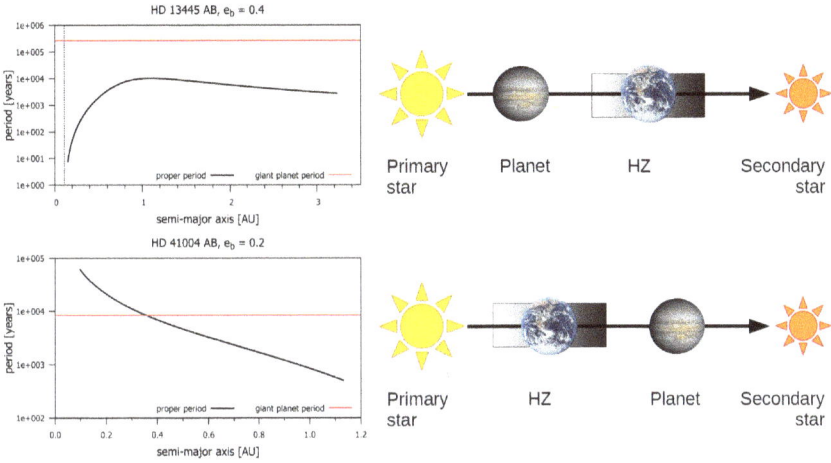

Figure 7.3: Giant planets interior (top) and exterior (bottom) to the HZ in binary star systems. The systems HD 13445 and HD 41004 point out the qualitatively different behavior of the secular period in the two cases. For details see the text.

the giant planet is located closer to the host star than the HZ, while the bottom panel shows the exterior case.[5] From the sample of S-type binary star systems in Figure 7.1, we can infer that the interior cases seem to be more numerous than the exterior cases. However, note that this prevalence of the interior case might be due to biased observations that tend to detect close-in planets. In effect, the semi-analytical method to find SRs works equally well for interior and exterior cases; nevertheless there are some caveats to it. For the interior cases, it is important to include the general relativistic precession, because for close-in planets it can be of a similar order of magnitude or even larger than the classical secular frequency. If this additional effect was neglected, it would turn out that generally interior planets cannot cause SR nearby or inside the HZ in binary star systems. This is illustrated in the upper panel of Figure 7.3 for the system HD 13445, for

---

[5] A third possible configuration arises when the giant planet itself is located exactly inside the HZ. We consider this as a special case and will not discuss it here, because it drastically lowers the probability of finding a terrestrial planet in the HZ.

which the proper secular periods of test planets (black curve) never approach the value of the giant planet (horizontal line). In contrast, for exterior planets we can always find an intersection of the two curves (bottom panel). This means that for this kind of configuration there is always a linear SR, and it is just a matter of the system's architecture whether or not this SR affects the HZ.

Finally, let us also discuss the limitations of the semi-analytical method. Although the method is fairly robust regarding the masses, eccentricities, and the semi-major axes of the planet and the secondary star, it can become difficult to obtain reliable estimates of the secular frequency for an inherently chaotic system. This is the case for large masses of the perturbing star and for very small stellar separations which put the planet close to the border of stable motion. Additionally, if the secular frequency happens to form an integer ratio with other frequencies (e.g., MMRs), then it becomes difficult to extract its correct value from the Fourier transform, and the location of the SR cannot be determined accurately.

7.4   Notes on Individual Systems

In the following, we briefly describe P-type and some of the tighter S-type systems (with $a_B \leq 500$ au) and collect the most important facts about them from the literature. Firstly, we summarize the system properties and in particular the results on habitable zones and resonances outlined in the previous sections.

Table 7.5 collects S-type systems from Table 7.1 with stellar separations up to 500 au. For these systems, first we calculated the region of stable motion, where planets of the host star remain on long-term stable orbits. This was done according to the methods presented in Section 2. Where possible we used the published value of the binary star's eccentricity, e_B (see Table 7.1); otherwise we assumed $e_B = 0.5$ as a typical value. The first column in Table 7.5 shows a_{crit}, which is a function of the stellar separation and mass-ratio. Since most of the planets are rather close to their host star, there are no issues regarding orbital stability, except for OGLE-2008-BLG-092L. In that system, the planet is reported at $a_P = 18$ au; hence it would be located beyond the stable region if the binary eccentricity was larger than $e_B > 0.1$. For this system, we can thus provide a dynamical constraint

Table 7.5: Summary of S-type systems and their important calculated parameters. The second column shows the region of stable motion around the host star (with border a_{crit}), the third column gives the extent of the classical habitable zone (for a single star) and the last column gives the location of a possible secular resonance.

System name	a_{crit} [au]	CHZ [au]	SR [au]
HD 109749	50.45	1.09–1.92	0.30–0.40
WASP-77	39.57	0.84–1.49	0.10–0.20
KELT-2	39.89	1.94–3.40	0.40–0.50
HD 114729	44.22	1.35–2.38	$\leq$0.1
Kepler-14	33.76	2.30–4.01	0.47–0.48
HD 27442	33.72	4.49–8.20	$\leq$0.1
TrES-2	32.33	0.97–1.72	0.13–0.15
HD 212301	36.00	1.05–1.84	0.17–0.18
HD 16141	34.29	0.89–1.58	3.20–3.22
HD 189733	34.28	0.57–1.04	0.25–0.26
WASP-85	25.31	0.98–1.71	0.28–0.29
GJ 15	21.83	0.15–0.28	0.33–0.34
HD 114762	22.01	1.23–2.17	–
WASP-2	14.65	0.65–1.18	0.22–0.24
HD 19994	21.32	1.95–3.42	–
GJ 3021	11.48	0.77–1.36	–
OGLE-2008-BLG-092L	8.78	0.39–0.73	16.0–16.8
τ Bootis	6.93	1.46–2.55	0.44–0.45
WASP-11/HAT-P-10	6.13	0.60–1.08	0.27–0.30
HD 126614	4.24	0.98–1.74	0.71–0.91
HD 41004 A	3.94	0.54–0.98	0.24–0.33
HD 41004 B	2.82	0.21–0.40	0.40–0.41
HD 196885	4.20	1.98–3.46	0.90–1.16
GJ 86	3.52	0.72–1.28	–
γ Cephei	3.84	3.40–6.19	0.63–0.82
OGLE-2013-BLG-0341 Bb	1.39	–	0.53–0.59
Kepler-420	1.02	1.00–1.77	–

for the binary eccentricity. The next column reports the distance range of the single star HZ around the host star. As demonstrated in this chapter (cf. Table 7.3), there is an excellent agreement between the SSHZ and the AHZ (see Chapter 6) for stars separated by 25 au or more. In one case (OGLE-2013-BLG-0341), the star's effective temperature is below the limit for the Kopparapu model, so for this system we do not show the HZ limits. Finally, the fourth column gives details about the probable location of a secular resonance. The SR location is calculated according to Section 7.3.1; an empty cell means that an SR is not possible in this system.

Apart from these derived quantities, a detailed stability analysis is necessary in the case of multi-planet binary star systems. Dynamical studies can also unravel other types of resonances (e.g., MMR) that have not been covered in this table.

7.4.1 *S-type systems*

HD 109749

A short-period planet (5.24 days) with about Saturn's mass was detected around HD 109749 by Fischer *et al.* (2006), an object with low chromospheric activity. The star HD 109749 B (SAO 223557), separated by about 8 arcsec, was used as a comparison star in the photometric analysis. Later, Desidera and Barbieri (2007) studied the stellar objects using astrometry and photometry, and concluded a physical connection. They derived stellar masses of 1.1 and 0.8 $M_\odot$, and the projected separation between the components of about 500 au.

HD 46375

Marcy *et al.* (2000) detected a Saturn-like mass companion around HD 46375 with a period of about 3 days using Doppler measurements. The search for photometric variations by transits failed (Henry, 2000), suggesting a lower line-of-sight inclination of the orbit. The star belongs to a visual pair separated by 10 arcsec; Mugrauer *et al.* (2006) confirmed physical binarity by both astrometry as well as photometry. They estimated a projected separation of about 350 au for the system and a mass ratio of 0.7, with both components showing sub-solar masses. Later, Gaulme *et al.* (2010) studied the planet host using precise CoRoT data and noted a possible detection of phase changes from the planet.

WASP-77

Maxted *et al.* (2013) reported the discovery of a transiting planet orbiting the brighter component of the visual binary star WASP-77. Gas planets as close to their host stars as WASP-77 Ab experience strong irradiation which can lead to strong planetary winds and noticeable mass loss of the planet. Salz *et al.* (2015) investigated the expansion of the planetary atmosphere by using Ly-α transit spectroscopy and found that WASP-77 Ab experiences very high mass loss rates. Evans *et al.* (2016, 2017) confirmed the planet and the reported orbital elements by direct imaging. Furthermore, Sirothia *et al.* (2014) reported the detection of WASP-77 in radio wavelengths.

KELT-2

Beatty *et al.* (2012) discovered a hot Jupiter transiting the bright primary star of the HD 42176 binary system. Recently, Martioli *et al.* (2018) confirmed the planet around KELT-2 A by high-precision near-infrared photometry.

HD 114729

A planet with less than a Jupiter-mass and a period of 1,135 days was found by Butler *et al.* (2003) using radial velocity data obtained during the Keck Precision Doppler Survey. The search for wide (sub)stellar companions of planet host stars by Mugrauer *et al.* (2005) resulted in the detection of a subsolar mass companion (0.25 $M_\odot$) at a projected distance of 282 au.

Kepler-14

Buchhave *et al.* (2011) discovered a hot Jupiter transiting an F-type star in a close visual binary with a period of 6.8 d. The components have comparable masses and are separated by only 0.3 arcsec, resulting in inaccurate parameters if the dilution of the host star's light remains uncorrected. The authors improved the analysis by a correction of the dilution. Southworth (2012) reinvestigated the system and obtained a somewhat lower mass for the planet host star, resulting also in a lower planet mass than listed in the detection paper. Huber *et al.* (2013) found good agreement with the results by Southworth (2012) using asteroseismic and spectroscopic analyses.

HD 27442

A planet around HD 27442 was detected by Butler *et al.* (2001) using Doppler measurements. It has an orbit similar to the Earth and has about 1.5 Jupiter-masses. HD 27442 is a wide binary system with a common proper motion, consisting of a subgiant (primary) and a relatively young, hot white dwarf companion (Mugrauer *et al.*, 2007a).

TrES-2

O'Donovan *et al.* (2006) detected a transiting hot Jupiter with a period of about 2.5 days around a solar-like star. The detection paper also confirms the planet by high-precision radial velocity measurements. This is the first planet detected in the Kepler field prior to that mission, and was designated later also as Kepler-1. Therefore, it generated a lot of attention, resulting in many detailed investigations about dynamics (e.g., Freistetter *et al.*, 2009) or transit timing analyses (e.g., Rabus *et al.*, 2009; Raetz *et al.*, 2014), but no additional planets were discovered. However, planetary emission from TrES-2 was noticed by O'Donovan *et al.* (2010) in Spitzer/IRAC data. The binary nature was studied by Daemgen *et al.* (2009) and they reported the detection of a faint companion with about half a solar mass.

HD 212301

The late F-type star HD 212301 hosts a hot Jupiter type planet, discovered by a radial velocity survey (Lo Curto *et al.*, 2006). A stellar co-moving companion of low mass (early M-type dwarf) was later identified by Mugrauer and Neuhäuser (2009) at a projected distance of 230 au.

HD 16141

A Saturn-mass companion around HD 16141, a nearby G-type star, was detected by Marcy *et al.* (2000). The planet orbits the star with a period of almost 76 days at 0.35 au. Mugrauer *et al.* (2005) discovered a stellar low-mass companion, later confirmed as an early M-type star (Mugrauer *et al.*, 2007b; Eggenberger *et al.*, 2007).

Kepler-410

Kepler-410 Ab was confirmed by Van Eylen *et al.* (2014) as a Neptune-sized exoplanet on an eccentric orbit around the bright star Kepler-410 A. By using asteroseismology, the transit light curve, adaptive optics, speckle images and Spitzer transit observations, it could be shown that the candidate can only be an exoplanet. Additionally, transit timing variations (TTV) indicated the presence of at least one additional (non-transiting) planet in the system. Recently, Gajdoš *et al.* (2017) presented a new analysis of the TTVs, where they analyze 70 transit times obtained by the Kepler satellite and found evidence for a second body orbiting Kepler-410 A with a period of approximately 970 days on a slightly eccentric orbit.

WASP-85

Brown (2015) reported the discovery of the transiting hot Jupiter exoplanet WASP-85 Ab using a combined analysis of spectroscopic and photometric data. Recently, Evans *et al.* (2017) determined more precise parameters for the binary orbit and found it to be moderately eccentric and inclined to the line of sight.

GJ 15

The low mass planet GJ 15 Ab ($M \sin i = 5.35 \pm 0.75\ M_{\oplus}$) orbiting an M dwarf was detected by Howard *et al.* (2014) using radial velocities from the Eta-Earth Survey (HIRES at Keck Observatory). Recently, Trifonov *et al.* (2017) combined CARMENES precise Doppler measurements with those available from HIRES and HARPS and re-investigated the system GJ 15. They found that neither the HIRES nor the CARMENES data show the previously announced signal, but found evidence for a possible long-period, Saturn-mass planet orbiting GJ 15 A.

HD 195019

Fischer *et al.* (1998) announced a Jupiter-mass planet orbiting the star HD 195019 with a period of 18.27 days. They were using Doppler measurements taken at Lick Observatory, with additional velocities from Keck Observatory. Vogt *et al.* (2000) provided updated orbital parameters for the planet of HD 195019.

WASP-2

Collier Cameron *et al.* (2007) detected a Jupiter-mass planet in the system WASP-2, using the SuperWASP wide-field transit survey and the radial velocity spectrograph SOPHIE at the Observatoire de Haute-Provence. Bergfors *et al.* (2013) announced the detection of a stellar companion in the system WASP-2. Wöllert *et al.* (2015) confirmed this stellar companion by using the Calar Alto 2.2 m telescope with the Lucky Imaging camera AstraLux Norte. Southworth *et al.* (2010) presented updated planetary parameters, using high-precision photometry of three transits and taking into account the light from this new discovered faint star. Becker *et al.* (2013) further improved the measurements of the hot Jupiter exoplanet WASP-2b and its orbital parameters, by using transit observations of the WASP-2 exoplanet system by the Apache Point Survey of Transit Lightcurves of Exoplanets (APOSTLE) program.

94 Ceti (HD 19994)

Mayor *et al.* (2004) announced a planet candidate in this system that was detected by radial-velocity (RV) measurements. The host star has a companion at a minimum separation of at least 100 au; we have adopted this minimum value for Table 7.1. This companion was first assumed to be an M-dwarf by Mayor *et al.* (2004), but Roell *et al.* (2012) argued that the companion is a close binary with a total mass of 0.9 $M_\odot$. The actual spectral types of the companions do not play a significant role for calculating the habitable zone boundaries, because of the large distance to the planet.

GJ 3021 (HD 1237)

The planet of GJ 3021 was announced by Naef *et al.* (2001), and was found with the RV technique. Mugrauer *et al.* (2007a) and Chauvin *et al.* (2007) found a long-period M-dwarf companion by follow-up observations with adaptive optics imaging and spectroscopy.

τ Bootis (HD 120136)

Butler *et al.* (1997) reported on the planetary companion of τ Boo, which is a hot Jupiter type object with <4 days orbital period. Eggenberger *et al.* (2004) attributed to this star a companion with a semi-major axis of ~240 au, while in Raghavan *et al.* (2006), the (projected) separation is given as 45 au.

This difference could be the result of a large eccentricity for the secondary star that is close to a periastron passage (Roberts *et al.*, 2011).

HD 126614

A long-term survey revealed that HD 126614 is a binary star system with a single giant planet (Howard *et al.*, 2010). The presence of the stellar companion was proposed to account for a linear trend in the RV signal; a direct observation with adaptive optics revealed the true nature of the system. As the secondary star was not known before, some of its orbital parameters are not well constrained; especially, the eccentricity is uncertain and just limited by $e_B < 0.6$. This eccentricity is the upper limit for the planet to be located in the dynamically stable zone of the host star. It is possible that there is a third star gravitationally bound to this system in a much wider orbit ($>1,000$ au).

HD 41004

HD 41004 consists of a K-type primary star and an M-type secondary star (Zucker *et al.*, 2004); both stars have substellar companions. The giant planet around HD 41004 A is listed as HD 41004 Ab; HD 41004 B has a close-in brown dwarf companion HD 41004 Bb with a minimum mass of 18 Jupiter-masses in a 1.3 day orbit. This system and its HZ have been investigated extensively in Pilat-Lohinger and Funk (2010), Funk *et al.* (2015) and Pilat-Lohinger *et al.* (2016).

HD 196885

Correia *et al.* (2008) first announced a planet orbiting the star HD 196885 A. They noted the presence of a stellar companion HD 196885 B, but they could only vaguely constrain its orbit. Chauvin *et al.* (2011) established the binary nature of the two objects and presented updated orbital elements from a combined astrometric and RV study for both the companion star and the planet. The inclination of the planet is still unknown, which motivated Giuppone *et al.* (2012) to perform a dynamical analysis of this system. Their results indicated that the giant planet is either nearly coplanar to the binary star's orbital plane, or conversely it must have a high inclination orbit ($i = 44$ deg prograde, or $i = 137$ deg retrograde). Note that both the giant planet and the secondary star have rather high eccentricities, and at apoastron the planet approaches very closely to the stability limit of this system.

γ Cephei (HD 222404)

γ Cephei was one of the first exoplanet candidates (Campbell *et al.*, 1988). The nature of the signal was disputed until Hatzes *et al.* (2003) convincingly demonstrated that it is indeed a planet. Neuhäuser *et al.* (2007) provided a new orbital solution and an update of the system parameters (e.g., semi-major axis and eccentricity), that were slightly different before. The host star is an evolved K1III giant, which pushes the habitable zone quite far away from the star. The dynamics of this system was studied by Dvorak *et al.* (2003), Pilat-Lohinger (2005), Haghighipour (2006), Giuppone *et al.* (2011) and Funk *et al.* (2015), among others.

Gliese 86 (HD 13445)

The planet of Gliese 86 was discovered by Queloz *et al.* (2000) following a RV measurement campaign. It was then already evident that the host star must possess another massive companion, which was characterized as a white dwarf by Mugrauer and Neuhäuser (2005) and Lagrange *et al.* (2006), while Fuhrmann *et al.* (2014) further constrained the parameters of this system. The HZ and dynamical stability of planets in this binary were investigated previously in Pilat-Lohinger and Funk (2010) and Funk *et al.* (2015).

Kepler-420 (KOI-1257)

The *Kepler* space-telescope planetary candidate KOI-1257 was found via transit photometry and confirmed by follow-up radial velocity observations (Santerne *et al.*, 2014). Due to the combination of both methods this tight binary system ($a_B = 5.3$ au) is rather well characterized. However, the objects in this system have large eccentricities, such that the presence of additional planets interior or exterior to the orbit of the discovered planet seems unlikely.

7.4.2   P-type systems

The following description collects the most important facts and data about the P-type systems in Table 7.2.

DP Leo

Qian *et al.* (2010) and Beuermann *et al.* (2011) investigated the system DP Leo using observational data from different sources. Both found that the transit timings can be explained by a planet orbiting the cataclysmic binary DP Leonis.

FW Tau, ROXs 42

Both planetary-mass companions ($M \approx M_{\mathrm{Jup}}$) were discovered by Kraus *et al.* (2014) in wide orbits around young stars by direct imaging and were confirmed by Bowler *et al.* (2014) using moderate-resolution, near-infrared integral field spectroscopy. Later Kraus *et al.* (2015) confirmed by using ALMA data that the faint companion to FW Tau hosts one of the least massive primordial disks known to date. They determined the mass of the disc, which is too low to build standard gas or ice giants, but could be large enough to allow the formation of a compact system of sub-Earth mass planets.

HU Aqr

Qian *et al.* (2011) announced the discovery of two circumbinary planets in the system HU Aqr, by using eclipse timing. In the same year, Horner *et al.* (2011) and Wittenmyer *et al.* (2012) showed that the suggested planetary system is highly unstable. Hinse *et al.* (2012) re-investigated the eclipse timing data and found a new orbital parameter solution for both planets, which leads to long-term orbital stability of the system. Later Goździewski *et al.* (2012) re-analyze the system, using additional precision light curves and an improved Keplerian, kinematic model of the light travel time effect. They found that the data are best explained by a single circumbinary companion. Bours *et al.* (2014) and Goździewski *et al.* (2015) re-analyze the system using new observational data and found a systematic deviation from the older ephemeris. This deviation cannot be explained by a single planet nor by a two-planet system. Therefore, further observations are needed to fully understand the system HU Aqr.

NN Ser

Beuermann *et al.* (2010) investigated the post-common envelope binary NN Ser and discovered a planetary system containing two planets in a 2:1 mean motion resonance. Using new observations, Beuermann *et al.* (2013) confirmed the two planet solution, where the planets have to be in the 2:1 mean motion resonance, since no non-resonant stable solutions could be found.

NY Vir

Qian *et al.* (2012b) discovered two circumbinary planets in the system NY Vir, by using eclipse timing variations. Later, Lee *et al.* (2014) investigated the system again and confirmed both planets.

OGLE-2007-BLG-349

OGLE-2007-BLG-349 is the first circumbinary planet found by a microlensing event. Bennett *et al.* (2016) investigated two kinds of models: 2-planet models with one star and circumbinary planet models. They found that only the circumbinary models are consistent with the HST data. These models indicated that the planet has a mass of $\approx 80 M_\oplus$ and is orbiting a pair of M-dwarfs, which makes this the lowest mass circumbinary planet system known.

PSR B1620-26 (WD J1623-266)

The existence of a planetary companion in the system PSR B1620-26 was first suspected by Backer *et al.* (1993). They found that a second, weakly bound companion object moving in a wide orbit around the main binary system explains best the deviation from the expected behavior. In the same year, Thorsett *et al.* (1993) described timing observations taken over a 5-year period and concluded that the data is consistent with the acceleration by a planet or an additional star. Arzoumanian *et al.* (1996) re-analyzed the system and found the planetary perturber hypothesis more reliable, whereas a stellar companion is still possible. Finally, Sigurdsson *et al.* (2003) found that the pulsar B1620-26 has two companions, one of stellar mass and one of planetary mass.

Ross 458 (DT Vir)

Goldman *et al.* (2010) searched for new cool brown dwarfs by using optical and near-infrared broadband photometry. In their investigation they found three brown dwarfs, among them is the circumbinary brown dwarf Ross 458 c.

RR Cae

A circumbinary planet in the short-period white-dwarf eclipsing binary system RR Cae was reported by Qian *et al.* (2012a). The planet was discovered using eclipse timing. In the data there is also evidence for another giant circumbinary planet in a wide orbit.

SR 12

SR 12 was discovered by Kuzuhara *et al.* (2011) using direct imaging; hence it is the first planetary-mass companion candidate directly imaged around a binary. Bowler *et al.* (2014) confirmed the substellar companion SR 12 C using near-infrared spectra.

Kepler planets:

Kepler-16

Slawson *et al.* (2011) provided an updated catalogue augmented with the second Kepler data release, including 2,165 eclipsing binaries found in the Kepler data. Additionally, they presented 4 eclipsing binaries that exhibit extra eclipse events (among them Kepler-16) and 8 systems that show clear eclipse timing variations. A detailed analysis of the Kepler data by Doyle *et al.* (2011) gave precise constraints on the absolute dimensions of all three bodies (both stars and the planet).

Kepler-34 and Kepler-35

Welsh *et al.* (2012) reported two transiting circumbinary planets: Kepler-34 b and Kepler-35 b. Kepler-34 b orbits two Sun-like stars every 289 days, whereas Kepler-35 b orbits a pair of lower mass stars (approximately 0.8 and 0.9 $M_\odot$) every 131 days. Both are low-density gas-giant planets with their orbits close to mean motion resonances. Chavez *et al.* (2015) found that Kepler-34 and possibly Kepler-35 demonstrate some features of resonance trapping.

Kepler-38

Orosz *et al.* (2012a) presented the discovery of the circumbinary planet Kepler-38 b, which orbits close to the stability limit.

Kepler-47

Orosz *et al.* (2012b) reported the detection of the system Kepler-47, consisting of two planets orbiting around an eclipsing pair of stars, which makes it the first circumbinary system with two known planets. This discovery established that close binary stars can host complete planetary systems. Kostov *et al.* (2013) confirmed the results of Orosz *et al.* (2012b).

Kepler-413

Kepler-413 b was discovered by Kostov *et al.* (2014). Because of its orbital precession, the inclination of the planet continuously changes and therefore the planet often fails to transit the primary star, which leads to large gaps in the light curves without any transits.

Kepler-451 (2M 1938+4603)

Kepler-451 b was discovered by Baran *et al.* (2015) using eclipse timing. In addition to the main signal, they discovered a long-term trend that may be an evolutionary effect, or a hint for more bodies. When assuming that the planet moves in the same orbital plane as the binary, it is a Jupiter-mass object, which makes it the lowest-massed object among all tertiary components detected in similar systems.

Kepler-453

Welsh *et al.* (2015) presented the discovery of Kepler-453 b, which orbits about an eclipsing binary well outside the dynamical instability zone. Furthermore it lies within the habitable zone of the binary.

Kepler-1647

Kostov *et al.* (2016) reported the discovery of a Kepler transiting circumbinary planet that has, unlike the previous discoveries, a long orbital period of 1,100 days. This makes Kepler-1647 b the longest period transiting circumbinary planet and also one of the largest period transiting planets. With a radius of $\approx 1\,R_{\mathrm{Jup}}$ it is also the largest circumbinary object known.

Chapter 8

Concluding Remarks

8.1 Do Binary Stars Provide Habitable Environments?

Planetary habitability is certainly in the spotlight of astrophysical research. In the case of binary star systems, in addition to the large number of processes that are important in single star systems, dynamical effects must be taken into account. First of all, different types of planetary motion have to be considered, where a planet orbits either one (circumstellar/S-type motion) or both stars (circumbinary/P-type motion). In both cases, orbital stability is limited to certain regions of the phase space. For S-type motion, the faraway secondary star sets an outer limit for stable planetary motion around the primary star. For P-type motion, the inner border of stable motion is essential since a certain distance from the two stars is required for long-term dynamical stability. In all cases, the stability borders are determined by the parameters (masses, distance and eccentricity) of the binary star system. It is obvious that the stable area around the primary star is smaller when the secondary star is more massive than the primary. Moreover, the smaller the distance of the two stars and the higher the binary orbital eccentricity, the smaller the stable area becomes. The extension of the stable area is important, because a necessary requirement for habitability in binary star systems is certainly that the habitable zone (HZ) is in the stable area.

Various studies have shown that stellar binarity might cause eccentric motion of a planet in the HZ so that the application of the classical HZ borders defined, for example by Kopparapu *et al.* (2014), is not advisable. Therefore, Eggl *et al.* (2012) defined three different HZs when taking into account the combined dynamical–radiative influence of the secondary star. Due to the planetary eccentricity we are led to distinguish between permanently HZ (PHZ) when the planet always moves in the HZ, the extended HZ (EHZ) when the eccentricities are moderate and the planet leaves the HZ only for a short time, and the averaged HZ (AHZ) when the planet can have a higher eccentricity and its orbit is only partly in the HZ. Comparisons of single and binary star HZs have shown that the AHZ corresponds quite well to the classical HZ of single stars. For the calculation of the different HZs, Eggl *et al.* (2012) developed an analytic method that does not need huge amounts of numerical computations.

In contrast to single stars, one can also recognize in binary star systems stronger perturbations at locations of mean motion resonances (MMRs), because of the secondary star's influence on the giant planet. In addition, secular resonances (SRs) can occur due to the combined perturbations of the giant planet and the secondary star. If a single giant planet orbits an isolated star — which is the case for many discovered exoplanetary systems — an SR will not be observed. We can observe a similar resonance feature only if two or more giant planets are orbiting a single star. The location of a resonance strongly depends on the parameters of the binary star system and the giant planet (i.e., mass, semi-major axis and eccentricity). A change of each parameter will modify the location of a resonance and, therefore, has a significant influence on the final architecture of a planetary system and on the habitability of a planet in the HZ. Thus it is important to know the locations of all resonances, which is easy to calculate for MMRs. For the SRs in binary star systems that contain a giant planet, we have presented a quick method (Pilat-Lohinger *et al.*, 2016; Bazsó *et al.*, 2017) to determine the SR's location without huge computational effort.

Resonances due to a giant planet also play an important role during terrestrial planet formation. A giant planet may restrict the area where terrestrial planets can form and may influence the shape of the orbits, in particular the eccentricity.

Even if terrestrial planets can be easily formed using N-body simulations, there are still improvements needed. In many cases, the applied "hit-and-stick" procedure for collisions does not reflect the reality. Further studies are needed to obtain better estimates for the merging of bodies and also for the water transport via planetesimals to protoplanets and planets. However, for the early phase of planet formation — which is concerned with the interplay of the binary stars with the protoplanets and the gaseous disc — we are faced with many open questions that still have to be solved. A first study in this context has been carried out by Gyergyovits *et al.* (2014), who developed a new hydrodynamics plus N-body code that runs on graphics processor cards for a better performance. However, such studies depend on many parameters that require further studies to shed more light on problems of planet formation in binary star systems. The detection of more than 100 planets in binary star systems[1] indicates that planet formation should be possible even in such perturbed environments. Dynamical studies provide methods that help to prove the stability of planetary motion and to determine the HZ and locations of perturbations in observed binary-star–giant planet systems and allow a preselection of systems having appropriate conditions for habitability from the dynamical point of view. The importance of such studies is obvious, as it is known from observations that more than 50% of the stars in the solar neighborhood form double or multiple star systems.

8.2   Will Solar System-like Configurations Help?

Even though we have evidence of more than 3,000 planets outside the Solar System, our Earth is still the only habitable planet we know so far. Therefore, the question arises whether we have to detect Solar System-like configurations to discover an exo-Earth. Various studies (Pilat-Lohinger *et al.*, 2008a,b; Pilat-Lohinger, 2015) addressed this question and investigated the HZ of Solar System-like configurations. In these investigations, they created such configurations by varying the semi-major axis of Saturn between 8 and 11 au and fixing Jupiter (and in Pilat-Lohinger *et al.* (2008b) also the other outer planets of the Solar System) to their known orbits. They showed that only Jupiter and Saturn dynamically influence the inner Solar System. Uranus and Neptune do not contribute significantly since they are not massive enough. In Pilat-Lohinger *et al.* (2008a), it

---

[1] see database at http://www.univie.ac.at/adg/schwarz/bincat_binary_star.html.

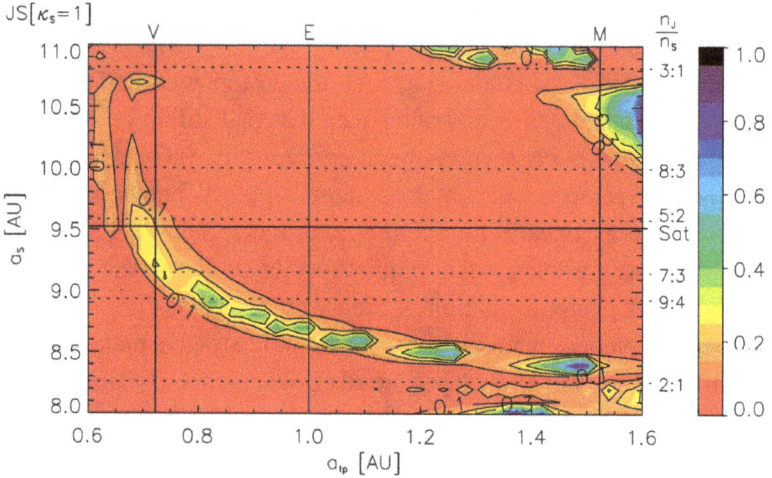

Figure 8.1: Stability map for test-planets perturbed by Jupiter and Saturn. The x-axis denotes the semi-major axes of the test-planets and the y-axis shows various initial semi-major axes of Saturn. Vertical black lines indicate the positions of the terrestrial planets Venus, Earth and Mars, respectively. The horizontal full line marks the current semi-major axis of Saturn, while the dashed lines show positions of MMRs. Different colors belong to different values of the maximum eccentricity that a test-planet achieved during the computations spanning 10^7 years. Details are described in the text. (This figure is reproduced by permission of the AAS from Pilat-Lohinger *et al.* (2008b).)

was shown that only a more massive Saturn or Uranus would change the dynamical behavior in the area of the terrestrial planets. Thus, we can reduce the system to Jupiter–Saturn-like configurations for studies regarding the habitability of Earth, unless we increase the masses of the other planets.

In Figure 8.1, we show the maximum eccentricities of test-planets between 0.6 and 1.6 au for different Jupiter–Saturn configurations. The red area indicates nearly circular motion within a computation time of 10^7 years, while other colors show areas of higher maximum eccentricities according to the color code. One can see that most of this region is not affected by the giant planets. Close to the 2:1 and 3:1 MMRs (indicated by the labels on the right side of the figure and the dashed horizontal lines) stronger perturbations are visible. The most significant perturbation is visualized by the arched band showing higher maximum eccentricity values; this marks a secular resonance with respect to the precession rate of Jupiter's perihelion, known as the ν_5 resonance in the Solar System. This figure shows three interesting features:

(1) Test planets have higher maximum eccentricities at Venus' position at about 0.72 au, which is due to the missing Earth in the computations. Pilat-Lohinger *et al.* (2008a) showed that adding the Earth to the system results in a completely different dynamical behavior of this region, where at Venus' position we would find nearly circular motion.
(2) Saturn at 8.7 au (note that the current position of Saturn is marked by the full horizontal line) would cause strong perturbations at Earth's position with variations in eccentricity up to 0.7, which would certainly change the conditions for Earth's habitability. The same behavior was found for a more massive Saturn (Pilat-Lohinger *et al.*, 2008b).
(3) The strong perturbations at the 2:1 MMR could have had an impact on the late stage of planet formation, in case of a resonance crossing as suggested by Tsiganis *et al.* (2005) in the NICE-Model, where especially the region of Mars would have been affected.

Considering Jupiter–Saturn-like configurations in binary star systems, it is necessary to address the question of stability. In tight binary star systems, such as γ Cephei, the Jupiter-like planet must be moved to 1.6 au for a stable configuration. For a G-type host-star, this corresponds to the outer region of the HZ, which probably prevents further planets from being in the HZ. In case of a stable Solar System-like Jupiter–Saturn configuration, the secondary star has to be at a distance > 100 au. At such a distance of the secondary star, Bancelin *et al.* (2016) showed that in combination with a Jupiter at 5.2 au, a secular resonance appears at about 1 au. Therefore, such systems provide interesting targets for further studies that are already in progress.

First results of an on-going study in this context are shown in Figure 8.2, where the stability of planetary motion in the HZ of tight binary star systems with stellar distances between 20 and 50 au has been investigated. For these systems, the locations of secular perturbations were determined by means of the newly developed method and are shown in Figure 8.2, where Jupiter and Saturn orbit the G-type host-star at 1.6 au and ~2.9 au, respectively. The system is perturbed by a secondary star at 30 au, for which the stellar type was varied between an M- and F-type. The application of the semi-analytical

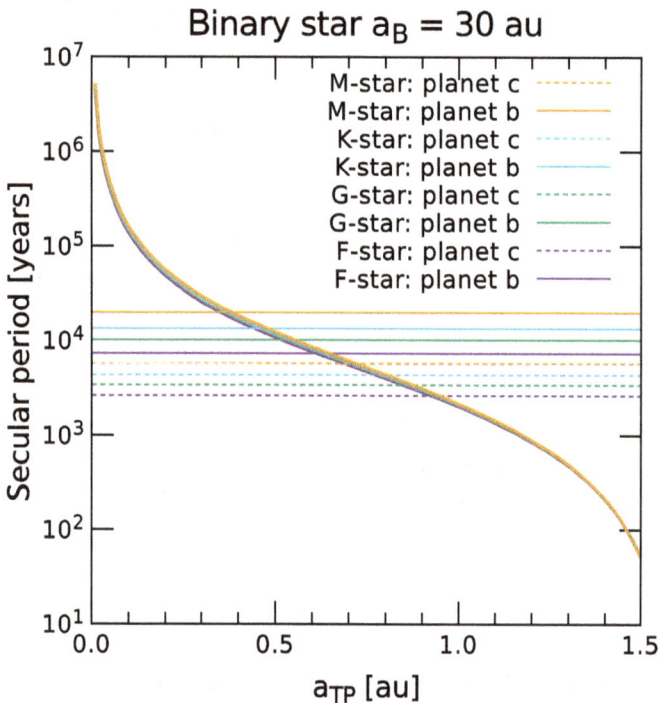

Figure 8.2: Locations of secular resonances for different binary configurations where the host-star is a G-type star and the secondary star is varied between M- and F-type stars at a fixed distance of 30 au. A Jupiter (labeled as planet b) at 1.6 au and Saturn (labeled as planet c) at 2.9 au orbit the G-type star. The horizontal full and dashed lines represent the proper orbital precession periods of the giant planets b and c, respectively. Intersections with the corresponding curve of proper periods of test-planets in this area show the locations of SRs.

method, which determines the location of a secular resonance (i.e., the intersection of a horizontal line with the respective curve of same color), shows that for all configurations the SR is at semi-major axes <1 au. This means that only the inner part of the HZ is perturbed. However, Figure 8.3 indicates for the G-G binary star system that there are additional strong perturbations due to MMRs with the giant planets (see the vertical stripes) and, therefore, it is very unlikely that such a system could host a terrestrial planet in the HZ.

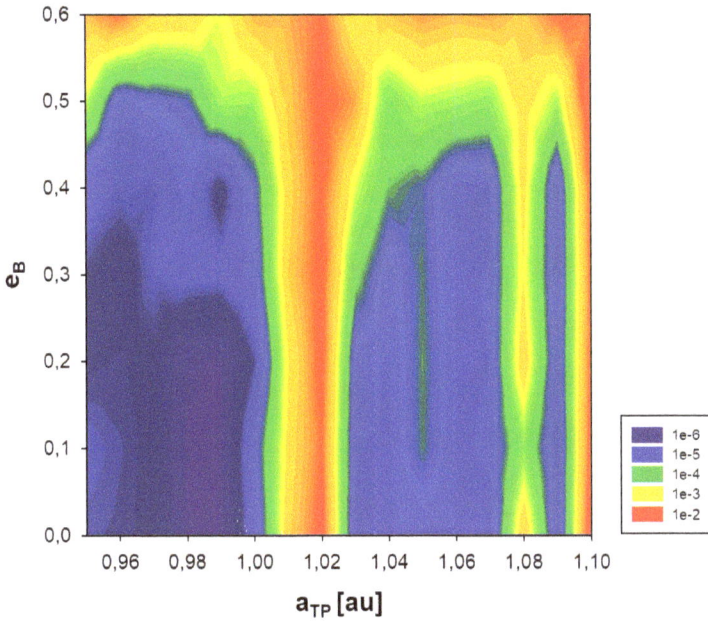

Figure 8.3: The FLI chaos indicator plot shows the stability of test-planets around the location of Earth in a G-G binary star system with stellar separation of 30 au, which also hosts a Jupiter–Saturn configuration. Jupiter orbits the host-star at 1.6 au and Saturn is at 2.9 au. According to the color code, stable areas are blue and chaotic motion is red.

Of course, "dynamical habitability" is only one of many conditions that need to be fulfilled. Long-term stability is certainly a basic requirement for a habitable planet, but it is well known that the proof of habitability of an exo-planet requires interdisciplinary studies that will engage the scientific community in the future.

Appendix

Glossary

A.1 Glossary of Abbreviations

Table A.1: List of abbreviations and definitions.

Acronym	Definition
AHZ	Averaged HZ; the long-term planetary insolation average must not exceed habitable limits; for planets with large climate inertia
CBHZ	Circumbinary HZ; one HZ around both stars in a P-type system
CME	Coronal Mass Ejection; a violent release of a large amount of charged particles and electromagnetic radiation into interplanetary space from a star's corona
CSHZ	Circumstellar HZ; a HZ around only one star of the S-type binary
EHZ	Extended HZ; 1-σ excursions of the planet's orbit beyond habitable insolation limits are permissible; for planets with intermediate climate inertia
EUV	Extreme Ultra-Violet; short wavelength radiation
FLI	Fast Lyapunov Indicator; a chaos indicator to quickly distinguish between regular and chaotic motion
GP	Giant Planet; a massive planet similar to Jupiter in the Solar System
GRP	General Relativistic Precession of the perihelion; a precession of the whole orbit induced by general relativistic effects in the vicinity of large masses
HZ	Habitable Zone; region around a star where a planet can sustain liquid water on its surface
HZc	Habitable Zone Crosser; an asteroid that evolves into an orbit that lets it traverse the HZ
IHZ	Isophote-based HZ; HZ borders are determined by insolation only
LCO	Lower Critical Orbit; limit radial distance up to which all starting positions are stable, it corresponds to a_{crit}
MMR	Mean Motion Resonance; resonance involving the orbital frequencies of two bodies

(Continued)

151

Table A.1: (*Continued*)

Acronym	Definition
MOID	Minimum Orbit Intersection Distance; the closest spatial distance between two Keplerian orbits
P-type	Planetary or circumbinary motion; a planet moving around the center of mass of a close binary star
PHZ	Permanent HZ; a planet on its evolving orbit cannot exceed habitable insolation limits; for planets with low climate inertia
RHZ	Radiative HZ; largest spherical shell to fit in the IHZ
RTBP	Restricted Three-Body Problem; a special case of the three-body problem where one of the masses is much smaller than the others and is negligible
S-type	Satellite or circumstellar motion; a planet moving around one component of a binary star system
SPH	Smoothed Particle Hydrodynamics; a mesh-free Lagrangian particle method for numerical hydrodynamic simulations of gases, fluids and solids
SR	Secular Resonance; resonance between the orbital precession frequencies of two bodies
SSHZ	Single Star HZ; classical HZ around a single star; no influence of a second radiation source
T-type	Trojan or libration motion; a planet moving in the vicinity of the Lagrangian equilibrium points
TBR	Three-Body Resonance; resonance that involves the orbital frequencies of three bodies
TP	Terrestrial Planet; a low-mass rocky planet similar to Earth
UCO	Upper Critical Orbit; limit radial distance beyond which all starting positions result in unstable orbits
WMF	Water Mass Fraction; the initial water content by mass of an asteroid
XUV	X-ray and Extreme Ultra-Violet; high energy electromagnetic radiation

A.2   Glossary of Symbols

Table A.2: List of variables and symbols.

Name	Dimension	Description
$\mathbb{A}$	[au^2]	spectrally weighted luminosity of Star A
a_B, a_P	[au]	semi-major axis of the double star/planetary orbit
a_{crit}	[au]	critical distance up to (from) which a planet's orbit is stable in S-type (P-type) motion
a_{GP}	[au]	semi-major axis of a giant planet
a_{TP}	[au]	semi-major axis of a test planet

(*Continued*)

Table A.2: (*Continued*)

Name	Dimension	Description
$AHZ(I, O)$	[au]	averaged HZ borders for S-type and P-type systems
au	[m]	astronomical unit of length defined as 149 597 870 700 meters
$\mathbb{B}$	[au^2]	spectrally weighted luminosity of Star B
b	[au]	$d/2$
d	[au]	star-to-star distance
e_B, e_P	[]	double star/planetary orbital eccentricity
e_{GP}, e_{TP}	[]	eccentricity of the giant planet/terrestrial planet
ϵ, η	[]	forced/free eccentricity of the planetary orbit
f_B, f_P	[rad]	true anomaly of the binary/planetary orbit
ϕ	[rad]	initial phase of the planetary orbit eccentricity vector
g	[rad day^{-1}]	secular frequency of the planetary orbit eccentricity
i	[rad]	orbital inclination
i_{TP}	[rad]	inclination of the terrestrial planet
$\mathbb{I}$	[]	momentary insolation function
$IHZ(I, O)$	[au]	isophote-based HZ borders
$\mathcal{G}$	[au^3 day^{-2} M$_\odot^{-1}$]	gravitational constant
l	[W]	stellar luminosity
L	[]	normalized stellar luminosity
λ	[rad]	mean longitude
$M_\odot$	[kg]	one solar mass; equivalent to 1.99×10^{30} kg
$M_\oplus$	[kg]	one Earth mass; equivalent to 5.97×10^{24} kg
m_A, m_B	[M$_\odot$]	stellar masses
μ	[]	mass ratio or reduced mass, $m_B/(m_A + m_B)$
n_B, n_P	[rad day^{-1}]	mean motion of the double star / planet
$PHZ(I, O)$	[au]	permanently HZ borders for S-type and P-type systems
ϖ_B, ϖ_P	[rad]	longitude of pericenter of the double star/ planetary orbit
Ω_P	[rad]	longitude of ascending node of the planetary orbit
Ψ	[rad]	relative position angle between binary star and planet
q_B, q_P	[au]	pericenter distance of the double star/planetary orbit
Q_B, Q_P	[au]	apocenter distance of the double star/planetary orbit
ρ	[au]	planet-star distance
r, r_A, r_B	[au]	planet-to-star distances
r_{AB}	[au]	distance of circumbinary planet to center of reference
$\vec{r}_{ji}$	[au]	relative position vector of two bodies
$RHZ(I, O)$	[au]	radiative HZ borders
s	[W m^{-2}]	insolation
S	[]	normalized insolation

(*Continued*)

Table A.2:   (*Continued*)

Name	Dimension	Description
$SAB_{I,O}$	[]	normalized insolation limits for inner and outer HZ borders of star A and B, spectral weights
T	[]	normalized effective temperature
T_{eff}	[K]	stellar effective temperature
$\bar{\theta}$	[rad]	impact angle in collision simulations
$\bar{v}$	[m s^{-1}]	impact speed in collision simulations
x, y, z	[au]	Cartesian coordinates with respect to the point of reference

Bibliography

Abe, Y., Ohtani, E., Okuchi, T., Righter, K., and Drake, M. (2000). Water in the early Earth, in R. M. Canup, K. Righter, *et al.* (eds.), *Origin of the Earth and Moon*, pp. 413–433.

Andrade-Ines, E., Beaugé, C., Michtchenko, T., and Robutel, P. (2016). Secular dynamics of S-type planetary orbits in binary star systems: Applicability domains of first- and second-order theories, *CMDA* **124**, pp. 405–432, doi:10.1007/s10569-015-9669-5, arXiv:1512.03585 [astro-ph.EP].

Andrade-Ines, E. and Eggl, S. (2017). Secular orbit evolution in systems with a strong external perturber — A simple and accurate model, *AJ* **153**, 148, doi:10.3847/1538-3881/153/4/148, arXiv:1701.03425 [astro-ph.EP].

Armstrong, D. J., Osborn, H. P., Brown, D. J. A., Faedi, F., Gómez Maqueo Chew, Y., Martin, D. V., Pollacco, D., and Udry, S. (2014). On the abundance of circumbinary planets, *MNRAS* **444**, pp. 1873–1883, doi:10.1093/mnras/stu1570, arXiv:1404.5617 [astro-ph.EP].

Artymowicz, P. and Lubow, S. H. (1994). Dynamics of binary-disk interaction. 1: Resonances and disk gap sizes, *ApJ* **421**, pp. 651–667, doi:10.1086/173679.

Arzoumanian, Z., Joshi, K., Rasio, F. A., and Thorsett, S. E. (1996). Orbital parameters of the PSR B1620-26 triple system, in S. Johnston, M. A. Walker, and M. Bailes (eds.), *IAU Colloq. 160: Pulsars: Problems and Progress*, *Astronomical Society of the Pacific Conference Series*, Vol. 105, pp. 525–530, astro-ph/9605141.

Backer, D. C., Foster, R. S., and Sallmen, S. (1993). A second companion of the millisecond pulsar 1620–26, *Nature* **365**, pp. 817–819, doi:10.1038/365817a0.

Bancelin, D., Hestroffer, D., and Thuillot, W. (2012). Numerical integration of dynamical systems with Lie series. Relativistic acceleration and non-gravitational forces, *CMDA* **112**, pp. 221–234, doi:10.1007/s10569-011-9393-8.

Bancelin, D., Pilat-Lohinger, E., and Bazsó, Á. (2016). Asteroid flux towards circumprimary habitable zones in binary star systems. II. Dynamics, *A&A* **591**, A120, doi:10.1051/0004-6361/201528035, arXiv:1512.08875 [astro-ph.EP].

Bancelin, D., Pilat-Lohinger, E., Eggl, S., Maindl, T. I., Schäfer, C., Speith, R., and Dvorak, R. (2015). Asteroid flux towards circumprimary habitable zones in binary star systems. I. Statistical overview, *A&A* **581**, A46, doi:10.1051/0004-6361/201526430, arXiv:1506.00993 [astro-ph.EP].

Bancelin, D., Pilat-Lohinger, E., Maindl, T. I., Ragossnig, F., and Schäfer, C. (2017). The influence of orbital resonances on the water transport to objects in the circumprimary habitable zone of binary star systems, *AJ* **153**, 269, doi:10.3847/1538-3881/aa7202, arXiv:1703.09450 [astro-ph.EP].

Baran, A. S., Zola, S., Blokesz, A., Østensen, R. H., and Silvotti, R. (2015). Detection of a planet in the sdB + M dwarf binary system 2M 1938+4603, *A&A* **577**, A146, doi:10.1051/0004-6361/201425392.

Barbieri, M., Marzari, F., and Scholl, H. (2002). Formation of terrestrial planets in close binary systems: The case of alpha Centauri A, *A&A* **396**, pp. 219–224, doi:10.1051/0004-6361:20021357, astro-ph/0209118.

Bazsó, Á., Pilat-Lohinger, E., Eggl, S., Funk, B., Bancelin, D., and Rau, G. (2017). Dynamics and habitability in circumstellar planetary systems of known binary stars, *MNRAS* **466**, pp. 1555–1566, doi:10.1093/mnras/stw3095, arXiv:1605.06769 [astro-ph.EP].

Beatty, T. G., Pepper, J., Siverd, R. J., Eastman, J. D., Bieryla, A., Latham, D. W., Buchhave, L. A., Jensen, E. L. N., Manner, M., Stassun, K. G., Gaudi, B. S., Berlind, P., Calkins, M. L., Collins, K., DePoy, D. L., Esquerdo, G. A., Fulton, B. J., Fűrész, G., Geary, J. C., Gould, A., Hebb, L., Kielkopf, J. F., Marshall, J. L., Pogge, R., Stanek, K. Z., Stefanik, R. P., Street, R., Szentgyorgyi, A. H., Trueblood, M., Trueblood, P., and Stutz, A. M. (2012). KELT-2Ab: A hot Jupiter transiting the bright (V = 8.77) primary star of a binary system, *ApJL* **756**, L39, doi:10.1088/2041-8205/756/2/L39, arXiv:1206.1592 [astro-ph.EP].

Becker, A. C., Kundurthy, P., Agol, E., Barnes, R., Williams, B. F., and Rose, A. E. (2013). Observations of the WASP-2 system by the APOSTLE program, *ApJL* **764**, L17, doi:10.1088/2041-8205/764/1/L17, arXiv:1301.3955 [astro-ph.EP].

Benest, D. (1988a). Planetary orbits in the elliptic restricted problem. I — The Alpha Centauri system, *A&A* **206**, pp. 143–146.

Benest, D. (1988b). Stable planetary orbits around one component in nearby binary stars, *Celestial Mechanics* **43**, pp. 47–53.

Benest, D. (1989). Planetary orbits in the elliptic restricted problem. II — The Sirius system, *A&A* **223**, pp. 361–364.

Benest, D. (1993). Stable planetary orbits around one component in nearby binary stars. II, *Celestial Mechanics and Dynamical Astronomy* **56**, pp. 45–50, doi:10.1007/BF00699718.

Benest, D. (1996). Planetary orbits in the elliptic restricted problem. III. The η Coronae Borealis system. *A&A* **314**, pp. 983–988.

Benest, D. (1998). Planetary orbits in the elliptic restricted problem. IV. The ADS 12033 system, *A&A* **332**, pp. 1147–1153.

Bennett, D. P., Rhie, S. H., Udalski, A., Gould, A., Tsapras, Y., Kubas, D., Bond, I. A., Greenhill, J., Cassan, A., Rattenbury, N. J., Boyajian, T. S., Luhn, J., Penny, M. T., Anderson, J., Abe, F., Bhattacharya, A., Botzler, C. S., Donachie, M., Freeman, M., Fukui, A., Hirao, Y., Itow, Y., Koshimoto, N., Li, M. C. A., Ling, C. H., Masuda, K., Matsubara, Y., Muraki, Y., Nagakane, M., Ohnishi, K., Oyokawa, H., Perrott, Y. C., Saito, T., Sharan, A., Sullivan, D. J., Sumi, T., Suzuki, D., Tristram, P. J., Yonehara, A., Yock, P. C. M., MOA Collaboration, Szymański, M. K., Soszyński, I., Ulaczyk, K., Wyrzykowski, Ł., OGLE Collaboration, Allen, W., DePoy, D., Gal-Yam, A., Gaudi, B. S., Han, C., Monard, I. A. G., Ofek, E.,

Pogge, R. W., μFUN Collaboration, Street, R. A., Bramich, D. M., Dominik, M., Horne, K., Snodgrass, C., Steele, I. A., Robonet Collaboration, Albrow, M. D., Bachelet, E., Batista, V., Beaulieu, J.-P., Brillant, S., Caldwell, J. A. R., Cole, A., Coutures, C., Dieters, S., Dominis Prester, D., Donatowicz, J., Fouqué, P., Hundertmark, M., Jørgensen, U. G., Kains, N., Kane, S. R., Marquette, J.-B., Menzies, J., Pollard, K. R., Ranc, C., Sahu, K. C., Wambsganss, J., Williams, A., Zub, M., and PLANET Collaboration (2016). The first circumbinary planet found by microlensing: OGLE-2007-BLG-349L(AB)c, *AJ* **152**, 125, doi:10.3847/0004-6256/152/5/125, arXiv:1609.06720 [astro-ph.EP].

Benz, W. and Asphaug, E. (1999). Catastrophic disruptions revisited, *Icarus* **142**, pp. 5–20, doi:10.1006/icar.1999.6204, astro-ph/9907117.

Benz, W., Mordasini, C., Alibert, Y., and Naef, D. (2008). Giant planet population synthesis: Comparing theory with observations, *Physica Scripta Volume T* **130**, 1, 014022, doi:10.1088/0031-8949/2008/T130/014022.

Bergfors, C., Brandner, W., Daemgen, S., Biller, B., Hippler, S., Janson, M., Kudryavtseva, N., Geißler, K., Henning, T., and Köhler, R. (2013). Stellar companions to exoplanet host stars: Lucky Imaging of transiting planet hosts, *MNRAS* **428**, pp. 182–189, doi:10.1093/mnras/sts019, arXiv:1209.4087 [astro-ph.SR].

Beuermann, K., Buhlmann, J., Diese, J., Dreizler, S., Hessman, F. V., Husser, T.-O., Miller, G. F., Nickol, N., Pons, R., Ruhr, D., Schmülling, H., Schwope, A. D., Sorge, T., Ulrichs, L., Winget, D. E., and Winget, K. I. (2011). The giant planet orbiting the cataclysmic binary DP Leonis, *A&A* **526**, A53, doi:10.1051/0004-6361/201015942, arXiv:1011.3905 [astro-ph.SR].

Beuermann, K., Dreizler, S., and Hessman, F. V. (2013). The quest for companions to post-common envelope binaries. IV. The 2:1 mean-motion resonance of the planets orbiting NN Serpentis, *A&A* **555**, A133, doi:10.1051/0004-6361/201220510, arXiv:1305.6494 [astro-ph.SR].

Beuermann, K., Hessman, F. V., Dreizler, S., Marsh, T. R., Parsons, S. G., Winget, D. E., Miller, G. F., Schreiber, M. R., Kley, W., Dhillon, V. S., Littlefair, S. P., Copperwheat, C. M., and Hermes, J. J. (2010). Two planets orbiting the recently formed post-common envelope binary NN Serpentis, *A&A* **521**, L60, doi:10.1051/0004-6361/201015472, arXiv:1010.3608 [astro-ph.SR].

Bolmont, E., Libert, A.-S., Leconte, J., and Selsis, F. (2016). Habitability of planets on eccentric orbits: Limits of the mean flux approximation, *A&A* **591**, A106, doi:10.1051/0004-6361/201628073, arXiv:1604.06091 [astro-ph.EP].

Boss, A. P. (2006). Gas giant protoplanets formed by disk instability in binary star systems, *ApJ* **641**, pp. 1148–1161, doi:10.1086/500530, astro-ph/0512477.

Bours, M. C. P., Marsh, T. R., Breedt, E., Copperwheat, C. M., Dhillon, V. S., Leckngam, A., Littlefair, S. P., Parsons, S. G., and Prasit, A. (2014). Testing the planetary models of HU Aquarii, *MNRAS* **445**, pp. 1924–1931, doi:10.1093/mnras/stu1879, arXiv:1409.3586 [astro-ph.SR].

Bowler, B. P., Liu, M. C., Kraus, A. L., and Mann, A. W. (2014). Spectroscopic confirmation of young planetary-mass companions on wide orbits, *ApJ* **784**, 65, doi:10.1088/0004-637X/784/1/65, arXiv:1401.7668 [astro-ph.EP].

Brasser, R., Morbidelli, A., Gomes, R., Tsiganis, K., and Levison, H. F. (2009). Constructing the secular architecture of the solar system II: The terrestrial planets, *A&A*

507, pp. 1053–1065, doi:10.1051/0004-6361/200912878, `arXiv:0909.1891` `[astro-ph.EP]`.

Bromley, B. C. and Kenyon, S. J. (2015). Planet formation around binary stars: Tatooine made easy, *ApJ* **806**, 98, doi:10.1088/0004-637X/806/1/98, `arXiv:1503.03876` `[astro-ph.EP]`.

Brown, D. J. A. (2015). Discovery of WASP-85 Ab: A hot Jupiter in a visual binary system, *European Planetary Science Congress* **10**, EPSC2015-603, `arXiv:1412.7761` `[astro-ph.EP]`.

Buchhave, L. A., Latham, D. W., Carter, J. A., Désert, J.-M., Torres, G., Adams, E. R., Bryson, S. T., Charbonneau, D. B., Ciardi, D. R., Kulesa, C., Dupree, A. K., Fischer, D. A., Fressin, F., Gautier, T. N., III, Gilliland, R. L., Howell, S. B., Isaacson, H., Jenkins, J. M., Marcy, G. W., McCarthy, D. W., Rowe, J. F., Batalha, N. M., Borucki, W. J., Brown, T. M., Caldwell, D. A., Christiansen, J. L., Cochran, W. D., Deming, D., Dunham, E. W., Everett, M., Ford, E. B., Fortney, J. J., Geary, J. C., Girouard, F. R., Haas, M. R., Holman, M. J., Horch, E., Klaus, T. C., Knutson, H. A., Koch, D. G., Kolodziejczak, J., Lissauer, J. J., Machalek, P., Mullally, F., Still, M. D., Quinn, S. N., Seager, S., Thompson, S. E., and Van Cleve, J. (2011). Kepler-14b: A massive hot Jupiter transiting an F Star in a close visual binary, *ApJS* **197**, 3, doi:10.1088/0067-0049/197/1/3, `arXiv:1106.5510` `[astro-ph.EP]`.

Bulirsch, R. and Stoer, J. (1966). Numerical treatment of ordinary differential equations by extrapolation methods, *Num. Math.* **8**, 1, pp. 1–13, doi:10.1007/BF02165234, `http://dx.doi.org/10.1007/BF02165234`.

Butler, R. P., Marcy, G. W., Vogt, S. S., Fischer, D. A., Henry, G. W., Laughlin, G., and Wright, J. T. (2003). Seven new Keck planets orbiting G and K dwarfs, *ApJ* **582**, pp. 455–466, doi:10.1086/344570.

Butler, R. P., Marcy, G. W., Williams, E., Hauser, H., and Shirts, P. (1997). Three new "51 Pegasi-Type" planets, *ApJL* **474**, pp. L115–L118, doi:10.1086/310444.

Butler, R. P., Tinney, C. G., Marcy, G. W., Jones, H. R. A., Penny, A. J., and Apps, K. (2001). Two new planets from the Anglo-Australian planet search, *ApJ* **555**, pp. 410–417, doi:10.1086/321467.

Campbell, B., Walker, G. A. H., and Yang, S. (1988). A search for substellar companions to solar-type stars, *ApJ* **331**, pp. 902–921, doi:10.1086/166608.

Chambers, J. E. (2010). N-body integrators for planets in binary star systems, in N. Haghighipour (ed.), *Planets in Binary Star Systems, Astrophysics and Space Science Library*, Vol. 366, p. 239, doi:10.1007/978-90-481-8687-7_9.

Chauvin, G., Beust, H., Lagrange, A.-M., and Eggenberger, A. (2011). Planetary systems in close binary stars: The case of HD 196885. Combined astrometric and radial velocity study, *A&A* **528**, A8, doi:10.1051/0004-6361/201015433.

Chauvin, G., Lagrange, A.-M., Udry, S., and Mayor, M. (2007). Characterization of the long-period companions of the exoplanet host stars: HD 196885, HD 1237 and HD 27442. VLT/NACO and SINFONI near-infrared, follow-up imaging and spectroscopy, *A&A* **475**, pp. 723–727, doi:10.1051/0004-6361:20067046.

Chavez, C. E., Georgakarakos, N., Prodan, S., Reyes-Ruiz, M., Aceves, H., Betancourt, F., and Perez-Tijerina, E. (2015). A dynamical stability study of Kepler Circumbinary planetary systems with one planet, *MNRAS* **446**, pp. 1283–1292, doi:10.1093/mnras/stu2142, `arXiv:1411.7761` `[astro-ph.EP]`.

Collier Cameron, A., Bouchy, F., Hébrard, G., Maxted, P., Pollacco, D., Pont, F., Skillen, I., Smalley, B., Street, R. A., West, R. G., Wilson, D. M., Aigrain, S., Christian, D. J., Clarkson, W. I., Enoch, B., Evans, A., Fitzsimmons, A., Fleenor, M., Gillon, M., Haswell, C. A., Hebb, L., Hellier, C., Hodgkin, S. T., Horne, K., Irwin, J., Kane, S. R., Keenan, F. P., Loeillet, B., Lister, T. A., Mayor, M., Moutou, C., Norton, A. J., Osborne, J., Parley, N., Queloz, D., Ryans, R., Triaud, A. H. M. J., Udry, S., and Wheatley, P. J. (2007). WASP-1b and WASP-2b: Two new transiting exoplanets detected with SuperWASP and SOPHIE, *MNRAS* **375**, pp. 951–957, doi:10.1111/j. 1365-2966.2006.11350.x, `astro-ph/0609688`.

Correia, A. C. M., Udry, S., Mayor, M., Eggenberger, A., Naef, D., Beuzit, J.-L., Perrier, C., Queloz, D., Sivan, J.-P., Pepe, F., Santos, N. C., and Ségransan, D. (2008). The ELODIE survey for northern extra-solar planets. IV. HD 196885, a close binary star with a 3.7-year planet, *A&A* **479**, pp. 271–275, doi:10.1051/0004-6361:20078908.

Cuntz, M. (2014). S-type and P-type habitability in stellar binary systems: A comprehensive approach. I. Method and applications, *ApJ* **780**, 14, doi:10.1088/0004-637X/780/1/ 14, `arXiv:1303.6645 [astro-ph.EP]`.

Cuntz, M. (2015). S-type and P-type habitability in stellar binary systems: A comprehensive approach. II. Elliptical orbits, *ApJ* **798**, 101, doi:10.1088/0004-637X/798/2/101, `arXiv:1409.3796 [astro-ph.SR]`.

Daemgen, S., Hormuth, F., Brandner, W., Bergfors, C., Janson, M., Hippler, S., and Henning, T. (2009). Binarity of transit host stars. Implications for planetary parameters, *A&A* **498**, pp. 567–574, doi:10.1051/0004-6361/200810988, `arXiv:0902.2179 [astro-ph.SR]`.

Desidera, S. and Barbieri, M. (2007). Properties of planets in binary systems. The role of binary separation, *A&A* **462**, pp. 345–353, doi:10.1051/0004-6361:20066319, `astro-ph/0610623`.

Doolin, S. and Blundell, K. M. (2011). The dynamics and stability of circumbinary orbits, *MNRAS* **418**, pp. 2656–2668, doi:10.1111/j.1365-2966.2011.19657.x, `arXiv:1108.4144 [astro-ph.SR]`.

Doyle, L. R., Carter, J. A., Fabrycky, D. C., Slawson, R. W., Howell, S. B., Winn, J. N., Orosz, J. A., Prsa, A., Welsh, W. F., Quinn, S. N., Latham, D., Torres, G., Buchhave, L. A., Marcy, G. W., Fortney, J. J., Shporer, A., Ford, E. B., Lissauer, J. J., Ragozzine, D., Rucker, M., Batalha, N., Jenkins, J. M., Borucki, W. J., Koch, D., Middour, C. K., Hall, J. R., McCauliff, S., Fanelli, M. N., Quintana, E. V., Holman, M. J., Caldwell, D. A., Still, M., Stefanik, R. P., Brown, W. R., Esquerdo, G. A., Tang, S., Furesz, G., Geary, J. C., Berlind, P., Calkins, M. L., Short, D. R., Steffen, J. H., Sasselov, D., Dunham, E. W., Cochran, W. D., Boss, A., Haas, M. R., Buzasi, D., and Fischer, D. (2011). Kepler-16: A transiting circumbinary planet, *Science* **333**, p. 1602, doi:10.1126/science.1210923, `arXiv:1109.3432 [astro-ph.EP]`.

Dressing, C. D., Spiegel, D. S., Scharf, C. A., Menou, K., and Raymond, S. N. (2010). Habitable climates: The influence of eccentricity, *ApJ* **721**, pp. 1295–1307, doi: 10.1088/0004-637X/721/2/1295, `arXiv:1002.4875 [astro-ph.EP]`.

Duchêne, G. and Kraus, A. (2013). Stellar multiplicity, *ARA&A* **51**, pp. 269–310, doi:10. 1146/annurev-astro-081710-102602, `arXiv:1303.3028 [astro-ph.SR]`.

Duquennoy, A., Mayor, M., and Halbwachs, J.-L. (1991). Multiplicity among solar type stars in the solar neighbourhood. I — CORAVEL radial velocity observations of 291 stars, *A&AS* **88**, pp. 281–324.

Dvorak, R. (1984). Numerical experiments on planetary orbits in double stars, *Celestial Mechanics* **34**, pp. 369–378, doi:10.1007/BF01235815.

Dvorak, R. (1986). Critical orbits in the elliptic restricted three-body problem, *A&A* **167**, pp. 379–386.

Dvorak, R., Froeschle, C., and Froeschle, C. (1989). Stability of outer planetary orbits (P-types) in binaries, *A&A* **226**, pp. 335–342.

Dvorak, R., Pilat-Lohinger, E., Funk, B., and Freistetter, F. (2003). Planets in habitable zones:. A study of the binary Gamma Cephei, *A&A* **398**, pp. L1–L4, doi:10.1051/ 0004-6361:20021805, astro-ph/0211289.

Eggenberger, A., Udry, S., Chauvin, G., Beuzit, J.-L., Lagrange, A.-M., Ségransan, D., and Mayor, M. (2007). The impact of stellar duplicity on planet occurrence and properties. I. Observational results of a VLT/NACO search for stellar companions to 130 nearby stars with and without planets, *A&A* **474**, pp. 273–291, doi:10.1051/ 0004-6361:20077447.

Eggenberger, A., Udry, S., and Mayor, M. (2004). Statistical properties of exoplanets. III. Planet properties and stellar multiplicity, *A&A* **417**, pp. 353–360, doi:10.1051/ 0004-6361:20034164, astro-ph/0402664.

Eggl, S. (2018). Habitable zones in binary star systems, in H. J. Deeg, J. A. Belmonte, V. Meadows, and R. Barnes (eds.), *Where Life May Arise: Habitability*, handbook of exoplanets (Springer International Publishing), ISBN 978-3-319-55332-0, http:// www.springer.com/de/book/9783319553320#aboutBook.

Eggl, S. and Dvorak, R. (2010). An introduction to common numerical integration codes used in dynamical astronomy, in J. Souchay and R. Dvorak (eds.), *Dynamics of Small Solar System Bodies and Exoplanets, Lecture Notes in Physics, Berlin Springer Verlag*, Vol. 790, pp. 431–480, doi:10.1007/978-3-642-04458-8_9.

Eggl, S., Haghighipour, N., and Pilat-Lohinger, E. (2013a). Detectability of earth-like planets in circumstellar habitable zones of binary star systems with sun-like components, *ApJ* **764**, 130, doi:10.1088/0004-637X/764/2/130, arXiv:1212.4884 [astro-ph.EP].

Eggl, S., Pilat-Lohinger, E., Funk, B., Georgakarakos, N., and Haghighipour, N. (2013b). Circumstellar habitable zones of binary-star systems in the solar neighbourhood, *MNRAS* **428**, pp. 3104–3113, doi:10.1093/mnras/sts257, arXiv:1210.5411 [astro-ph.EP].

Eggl, S., Pilat-Lohinger, E., Georgakarakos, N., Gyergyovits, M., and Funk, B. (2012). An analytic method to determine habitable zones for S-type planetary orbits in binary star systems, *ApJ* **752**, 74, doi:10.1088/0004-637X/752/1/74, arXiv:1204.2496 [astro-ph.EP].

Ellis, K. M. and Murray, C. D. (2000). The disturbing function in solar system dynamics, *Icarus* **147**, pp. 129–144, doi:10.1006/icar.2000.6399.

Endl, M., Bergmann, C., Hearnshaw, J., Barnes, S. I., Wittenmyer, R. A., Ramm, D., Kilmartin, P., Gunn, F., and Brogt, E. (2015). The Mt John University Observatory search for Earth-mass planets in the habitable zone of α Centauri, *International*

Journal of Astrobiology **14**, pp. 305–312, doi:10.1017/S1473550414000081, arXiv:1403.4809 [astro-ph.EP].

Evans, D. F., Southworth, J., Maxted, P. F. L., Skottfelt, J., Hundertmark, M., Jørgensen, U. G., Dominik, M., Alsubai, K. A., Andersen, M. I., Bozza, V., Bramich, D. M., Burgdorf, M. J., Ciceri, S., D'Ago, G., Figuera Jaimes, R., Gu, S.-H., Haugbølle, T., Hinse, T. C., Juncher, D., Kains, N., Kerins, E., Korhonen, H., Kuffmeier, M., Mancini, L., Peixinho, N., Popovas, A., Rabus, M., Rahvar, S., Schmidt, R. W., Snodgrass, C., Starkey, D., Surdej, J., Tronsgaard, R., von Essen, C., Wang, Y.-B., and Wertz, O. (2016). High-resolution Imaging of Transiting Extrasolar Planetary systems (HITEP). I. Lucky imaging observations of 101 systems in the southern hemisphere, *A&A* **589**, A58, doi:10.1051/0004-6361/201527970, arXiv:1603.03274 [astro-ph.EP].

Evans, D. F., Southworth, J., Smalley, B., Jørgensen, U. G., Dominik, M., Andersen, M. I., Bozza, V., Bramich, D. M., Burgdorf, M. J., Ciceri, S., D'Ago, G., Figuera Jaimes, R., Gu, S.-H., Hinse, T. C., Henning, T., Hundertmark, M., Kains, N., Kerins, E., Korhonen, H., Kokotanekova, R., Kuffmeier, M., Longa-Peña, P., Mancini, L., MacKenzie, J., Popovas, A., Rabus, M., Rahvar, S., Sajadian, S., Snodgrass, C., Skottfelt, J., Surdej, J., Tronsgaard, R., Unda-Sanzana, E., von Essen, C., Wang, Y.-B., and Wertz, O. (2017). High-resolution Imaging of Transiting Extrasolar Planetary systems (HITEP). II. Lucky Imaging results from 2015 and 2016, *ArXiv e-prints* arXiv:1709.07476 [astro-ph.EP].

Everhart, E. (1974). Implicit single-sequence methods for integrating orbits, *Cel. Mech.* **10**, pp. 35–55, doi:10.1007/BF01261877.

Fischer, D. A., Butler, R. P., Marcy, G. W., and Vogt, S. S. (1998). The lick observatory planet search, in *American Astronomical Society Meeting Abstracts, Bulletin of the American Astronomical Society*, Vol. 30, p. 1391.

Fischer, D. A., Laughlin, G., Marcy, G. W., Butler, R. P., Vogt, S. S., Johnson, J. A., Henry, G. W., McCarthy, C., Ammons, M., Robinson, S., Strader, J., Valenti, J. A., McCullough, P. R., Charbonneau, D., Haislip, J., Knutson, H. A., Reichart, D. E., McGee, P., Monard, B., Wright, J. T., Ida, S., Sato, B., and Minniti, D. (2006). The N2K consortium. III. Short-period planets orbiting HD 149143 and HD 109749, *ApJ* **637**, pp. 1094–1101, doi:10.1086/498557.

Forgan, D. (2014). Assessing circumbinary habitable zones using latitudinal energy balance modeling, *MNRAS* **437**, pp. 1352–1361, doi:10.1093/mnras/stt1964, arXiv:1310.3611 [astro-ph.EP].

Forgan, D. (2016). Milankovitch cycles of terrestrial planets in binary star systems, *MNRAS* **463**, pp. 2768–2780, doi:10.1093/mnras/stw2098, arXiv:1608.05592 [astro-ph.EP].

Forgan, D. H., Mead, A., Cockell, C. S., and Raven, J. A. (2015). Surface flux patterns on planets in circumbinary systems and potential for photosynthesis, *International Journal of Astrobiology* **14**, pp. 465–478, doi:10.1017/S147355041400041X, arXiv:1408.5277 [astro-ph.EP].

Freistetter, F., Süli, Á., and Funk, B. (2009). Dynamics of the TrES-2 system, *Astronomische Nachrichten* **330**, pp. 469–474, doi:10.1002/asna.200811199, arXiv:0905.1806 [astro-ph.EP].

Frigo, M. and Johnson, S. G. (2005). The design and implementation of FFTW3, *Proceedings of the IEEE* **93**, 2, pp. 216–231, special issue on "Program Generation, Optimization, and Platform Adaptation".

Froeschlé, C., Lega, E., and Gonczi, R. (1997). Fast Lyapunov indicators. Application to asteroidal motion, *Celestial Mechanics and Dynamical Astronomy* **67**, pp. 41–62, doi:10.1023/A:1008276418601.

Froeschlé, C. and Scholl, H. (1989). The three principal secular resonances nu(5), nu(6), and nu(16) in the asteroidal belt, *CMDA* **46**, pp. 231–251, doi:10.1007/BF00049260.

Fuhrmann, K., Chini, R., Buda, L.-S., and Pozo Nuñez, F. (2014). On the age of Gliese 86, *ApJ* **785**, 68, doi:10.1088/0004-637X/785/1/68.

Funk, B., Pilat-Lohinger, E., and Eggl, S. (2015). Can there be additional rocky planets in the Habitable Zone of tight binary stars with a known gas giant? *MNRAS* **448**, pp. 3797–3805, doi:10.1093/mnras/stv253, arXiv:1505.07069 [astro-ph.EP].

Gajdoš, P., Parimucha, Š., Hambálek, Ľ., and Vaňko, M. (2017). Transit-timing variations in the system Kepler-410Ab, *MNRAS* **469**, pp. 2907–2912, doi:10.1093/mnras/stx963, arXiv:1704.05663 [astro-ph.SR].

Gallardo, T. (2006). Atlas of the mean motion resonances in the Solar System, *Icarus* **184**, pp. 29–38, doi:10.1016/j.icarus.2006.04.001.

Gaulme, P., Vannier, M., Guillot, T., Mosser, B., Mary, D., Weiss, W. W., Schmider, F.-X., Bourguignon, S., Deeg, H. J., Régulo, C., Aigrain, S., Schneider, J., Bruntt, H., Deheuvels, S., Donati, J.-F., Appourchaux, T., Auvergne, M., Baglin, A., Baudin, F., Catala, C., Michel, E., and Samadi, R. (2010). Possible detection of phase changes from the non-transiting planet HD 46375b by CoRoT, *A&A* **518**, L153, doi:10.1051/0004-6361/201014303, arXiv:1011.2690 [astro-ph.EP].

Georgakarakos, N. (2003). Eccentricity evolution in hierarchical triple systems with eccentric outer binaries, *MNRAS* **345**, pp. 340–348, doi:10.1046/j.1365-8711.2003.06942.x, arXiv:1408.5890 [astro-ph.EP].

Georgakarakos, N. (2013). The dependence of the stability of hierarchical triple systems on the orbital inclination, *New Astronomy* **23**, pp. 41–48, arXiv:1302.5599 [astro-ph.EP].

Georgakarakos, N. and Eggl, S. (2015). Analytic orbit propagation for transiting circumbinary planets, *ApJ* **802**, 94, doi:10.1088/0004-637X/802/2/94, arXiv:1502.06387 [astro-ph.EP].

Giuppone, C. A., Leiva, A. M., Correa-Otto, J., and Beaugé, C. (2011). Secular dynamics of planetesimals in tight binary systems: Application to γ-Cephei, *A&A* **530**, A103, doi:10.1051/0004-6361/201016375, arXiv:1105.0243 [astro-ph.EP].

Giuppone, C. A., Morais, M. H. M., Boué, G., and Correia, A. C. M. (2012). Dynamical analysis and constraints for the HD 196885 system, *A&A* **541**, A151, doi:10.1051/0004-6361/201118356, arXiv:1203.5249 [astro-ph.EP].

Gladman, B. J., Migliorini, F., Morbidelli, A., Zappala, V., Michel, P., Cellino, A., Froeschle, C., Levison, H. F., Bailey, M., and Duncan, M. (1997). Dynamical lifetimes of objects injected into asteroid belt resonances, *Science* **277**, pp. 197–201.

Godolt, M., Grenfell, J. L., Kitzmann, D., Kunze, M., Langematz, U., Patzer, A. B. C., Rauer, H., and Stracke, B. (2016). Assessing the habitability of planets with Earth-like atmospheres with 1D and 3D climate modeling, *A&A* **592**, A36, doi:10.1051/0004-6361/201628413, arXiv:1605.08231 [astro-ph.EP].

Goldman, B., Marsat, S., Henning, T., Clemens, C., and Greiner, J. (2010). A new benchmark T8-9 brown dwarf and a couple of new mid-T dwarfs from the UKIDSS DR5+ LAS, *MNRAS* **405**, pp. 1140–1152, doi:10.1111/j.1365-2966.2010.16524.x, arXiv:1002.2637 [astro-ph.SR].

Goździewski, K., Nasiroglu, I., Słowikowska, A., Beuermann, K., Kanbach, G., Gauza, B., Maciejewski, A. J., Schwarz, R., Schwope, A. D., Hinse, T. C., Haghighipour, N., Burwitz, V., Słonina, M., and Rau, A. (2012). On the HU Aquarii planetary system hypothesis, *MNRAS* **425**, pp. 930–949, doi:10.1111/j.1365-2966.2012.21341. x, arXiv:1205.4164 [astro-ph.EP].

Goździewski, K., Słowikowska, A., Dimitrov, D., Krzeszowski, K., Żejmo, M., Kanbach, G., Burwitz, V., Rau, A., Irawati, P., Richichi, A., Gawroński, M., Nowak, G., Nasiroglu, I., and Kubicki, D. (2015). The HU Aqr planetary system hypothesis revisited, *MNRAS* **448**, pp. 1118–1136, doi:10.1093/mnras/stu2728, arXiv:1412.5899 [astro-ph.EP].

Güdel, M. (2007). The Sun in time: Activity and environment, *Living Reviews in Solar Physics* **4**, 3, doi:10.12942/lrsp-2007-3, arXiv:0712.1763.

Güdel, M., Dvorak, R., Erkaev, N., Kasting, J., Khodachenko, M., Lammer, H., Pilat-Lohinger, E., Rauer, H., Ribas, I., and Wood, B. E. (2014). Astrophysical conditions for planetary habitability, *Protostars and Planets VI* , pp. 883–906doi:10.2458/azu_uapress_9780816531240-ch038, arXiv:1407.8174 [astro-ph.EP].

Guedes, J. M., Rivera, E. J., Davis, E., Laughlin, G., Quintana, E. V., and Fischer, D. A. (2008). Formation and detectability of terrestrial planets around α Centauri B, *ApJ* **679**, 1582-1587, doi:10.1086/587799, arXiv:0802.3482.

Gyergyovits, M., Eggl, S., Pilat-Lohinger, E., and Theis, C. (2014). Disc-protoplanet interaction. Influence of circumprimary radiative discs on self-gravitating protoplanetary bodies in binary star systems, *A&A* **566**, A114, doi:10.1051/0004-6361/201321854, arXiv:1405.5056 [astro-ph.EP].

Haghighipour, N. (2006). Dynamical stability and habitability of the γ Cephei binary-planetary system, *ApJ* **644**, pp. 543–550, doi:10.1086/503351, astro-ph/0509659.

Haghighipour, N. and Kaltenegger, L. (2013). Calculating the habitable zone of binary star systems. II. P-type binaries, *ApJ* **777**, 166, doi:10.1088/0004-637X/777/2/166, arXiv:1306.2890 [astro-ph.EP].

Haghighipour, N. and Raymond, S. N. (2007). Habitable planet formation in binary planetary systems, *ApJ* **666**, pp. 436–446, doi:10.1086/520501, astro-ph/0702706.

Halbwachs, J. L., Mayor, M., Udry, S., and Arenou, F. (2003). Multiplicity among solar-type stars. III. Statistical properties of the F7-K binaries with periods up to 10 years, *A&A* **397**, pp. 159–175, doi:10.1051/0004-6361:20021507.

Han, E., Wang, S. X., Wright, J. T., Feng, Y. K., Zhao, M., Fakhouri, O., Brown, J. I., and Hancock, C. (2014). Exoplanet orbit database. II. Updates to Exoplanets.org, *PASP* **126**, pp. 827–837, doi:10.1086/678447, arXiv:1409.7709 [astro-ph.EP].

Hanslmeier, A. (ed.) (2007). *The Sun and Space Weather, Astrophysics and Space Science Library*, Vol. 347, doi:10.1007/978-1-4020-5604-8.

Hanslmeier, A. and Dvorak, R. (1984). Numerical integration with Lie series, *A&A* **132**, p. 203.

Harrington, R. S. (1977). Planetary orbits in binary stars, *AJ* **82**, pp. 753–756, doi:10.1086/112121.

Hart, M. H. (1978). The evolution of the atmosphere of the earth, *Icarus* **33**, pp. 23–39, doi:10.1016/0019-1035(78)90021-0.

Hart, M. H. (1979). Habitable zones about main sequence stars, *Icarus* **37**, pp. 351–357, doi:10.1016/0019-1035(79)90141-6.

Hatzes, A. P., Cochran, W. D., Endl, M., McArthur, B., Paulson, D. B., Walker, G. A. H., Campbell, B., and Yang, S. (2003). A planetary companion to γ Cephei A, *ApJ* **599**, pp. 1383–1394, doi:10.1086/379281, arXiv:astro-ph/0305110.

Hayashi, C., Nakazawa, K., and Mizuno, H. (1979). Earth's melting due to the blanketing effect of the primordial dense atmosphere, *Earth and Planetary Science Letters* **43**, pp. 22–28, doi:10.1016/0012-821X(79)90152-3.

Heintz, W. D. (1978). Double stars /Revised edition/, *Geophysics and Astrophysics Monographs* **15**.

Henry, G. W. (2000). Search for transits of a short-period, sub-Saturn extrasolar planet orbiting HD 46375, *ApJL* **536**, pp. L47–L48, doi:10.1086/312726.

Heppenheimer, T. A. (1978). On the formation of planets in binary star systems, *A&A* **65**, pp. 421–426.

Hinse, T. C., Lee, J. W., Goździewski, K., Haghighipour, N., Lee, C.-U., and Scullion, E. M. (2012). New light-travel time models and orbital stability study of the proposed planetary system HU Aquarii, *MNRAS* **420**, pp. 3609–3620, doi:10.1111/j.1365-2966.2011.20283.x, arXiv:1112.0066 [astro-ph.EP].

Holman, M. J. and Wiegert, P. A. (1999). Long-term stability of planets in binary systems, *AJ* **117**, pp. 621–628, doi:10.1086/300695, astro-ph/9809315.

Horner, J., Marshall, J. P., Wittenmyer, R. A., and Tinney, C. G. (2011). A dynamical analysis of the proposed HU Aquarii planetary system, *MNRAS* **416**, pp. L11–L15, doi:10.1111/j.1745-3933.2011.01087.x, arXiv:1106.0777 [astro-ph.EP].

Howard, A. W., Johnson, J. A., Marcy, G. W., Fischer, D. A., Wright, J. T., Bernat, D., Henry, G. W., Peek, K. M. G., Isaacson, H., Apps, K., Endl, M., Cochran, W. D., Valenti, J. A., Anderson, J., and Piskunov, N. E. (2010). The California planet survey. I. Four new giant exoplanets, *ApJ* **721**, pp. 1467–1481, doi:10.1088/0004-637X/721/2/1467, arXiv:1003.3488 [astro-ph.EP].

Howard, A. W., Marcy, G. W., Fischer, D. A., Isaacson, H., Muirhead, P. S., Henry, G. W., Boyajian, T. S., von Braun, K., Becker, J. C., Wright, J. T., and Johnson, J. A. (2014). The NASA-UC-UH ETA-Earth program. IV. A low-mass planet orbiting an M dwarf 3.6 PC from Earth, *ApJ* **794**, 51, doi:10.1088/0004-637X/794/1/51, arXiv:1408.5645 [astro-ph.EP].

Huang, S.-S. (1959). Occurrence of life in the universe, *PASP* **47**, pp. 397–402.

Huang, S.-S. (1960). Life-supporting regions in the vicinity of binary systems, *PASP* **72**, p. 106, doi:10.1086/127489.

Huber, D., Chaplin, W. J., Christensen-Dalsgaard, J., Gilliland, R. L., Kjeldsen, H., Buchhave, L. A., Fischer, D. A., Lissauer, J. J., Rowe, J. F., Sanchis-Ojeda, R., Basu, S., Handberg, R., Hekker, S., Howard, A. W., Isaacson, H., Karoff, C., Latham, D. W., Lund, M. N., Lundkvist, M., Marcy, G. W., Miglio, A., Silva Aguirre, V., Stello, D., Arentoft, T., Barclay, T., Bedding, T. R., Burke, C. J., Christiansen, J. L., Elsworth, Y. P., Haas, M. R., Kawaler, S. D., Metcalfe, T. S., Mullally, F., and Thompson, S. E. (2013). Fundamental properties of Kepler planet-candidate host stars using asteroseismology, *ApJ* **767**, 127, doi:10.1088/0004-637X/767/2/127, arXiv:1302.2624 [astro-ph.SR].

Izidoro, A., de Souza Torres, K., Winter, O. C., and Haghighipour, N. (2013). A compound model for the origin of earth's water, *ApJ* **767**, 54, doi:10.1088/0004-637X/767/1/54, arXiv:1302.1233 [astro-ph.EP].

Jaime, L. G., Pichardo, B., and Aguilar, L. (2012). Regions of dynamical stability for discs and planets in binary stars of the solar neighboring, *MNRAS* **427**, pp. 2723–2733, doi: 10.1111/j.1365-2966.2012.21839.x, arXiv:1208.2051 [astro-ph.EP].

Jang-Condell, H. (2015). On the likelihood of planet formation in close binaries, *ApJ* **799**, 147, doi:10.1088/0004-637X/799/2/147, arXiv:1501.00617 [astro-ph.EP].

Johnstone, C. P., Güdel, M., Brott, I., and Lüftinger, T. (2015a). Stellar winds on the main-sequence. II. The evolution of rotation and winds, *A&A* **577**, A28, doi:10.1051/0004-6361/201425301, arXiv:1503.07494 [astro-ph.SR].

Johnstone, C. P., Güdel, M., Lüftinger, T., Toth, G., and Brott, I. (2015b). Stellar winds on the main-sequence. I. Wind model, *A&A* **577**, A27, doi:10.1051/0004-6361/201425300, arXiv:1503.06669 [astro-ph.SR].

Johnstone, C. P., Zhilkin, A., Pilat-Lohinger, E., Bisikalo, D., Güdel, M., and Eggl, S. (2015c). Colliding winds in low-mass binary star systems: Wind interactions and implications for habitable planets, *A&A* **577**, A122, doi:10.1051/0004-6361/201425134, arXiv:1502.03334 [astro-ph.SR].

Kaib, N. A., Raymond, S. N., and Duncan, M. (2013). Planetary system disruption by Galactic perturbations to wide binary stars, *Nature* **493**, pp. 381–384, doi:10.1038/nature11780, arXiv:1301.3145 [astro-ph.EP].

Kaltenegger, L. and Haghighipour, N. (2013). Calculating the habitable zone of binary star systems. I. S-type binaries, *ApJ* **777**, 165, doi:10.1088/0004-637X/777/2/165, arXiv:1306.2889 [astro-ph.EP].

Kane, S. R. and Hinkel, N. R. (2013). On the habitable zones of circumbinary planetary systems, *ApJ* **762**, 7, doi:10.1088/0004-637X/762/1/7, arXiv:1211.2812 [astro-ph.EP].

Kasting, J. F. (1988). Runaway and moist greenhouse atmospheres and the evolution of earth and Venus, *Icarus* **74**, pp. 472–494, doi:10.1016/0019-1035(88)90116-9.

Kasting, J. F., Whitmire, D. P., and Reynolds, R. T. (1993). Habitable zones around main sequence stars, *Icarus* **101**, pp. 108–128, doi:10.1006/icar.1993.1010.

Kervella, P., Thévenin, F., Ségransan, D., Berthomieu, G., Lopez, B., Morel, P., and Provost, J. (2003). The diameters of alpha Centauri A and B. A comparison of the asteroseismic and VINCI/VLTI views, *A&A* **404**, pp. 1087–1097, doi:10.1051/0004-6361:20030570, astro-ph/0303634.

Khodachenko, M. L., Alexeev, I., Belenkaya, E., Lammer, H., Grießmeier, J.-M., Leitzinger, M., Odert, P., Zaqarashvili, T., and Rucker, H. O. (2012). Magnetospheres of "hot Jupiters": The importance of magnetodisks in shaping a magnetospheric obstacle, *ApJ* **744**, 70, doi:10.1088/0004-637X/744/1/70.

Khodachenko, M. L., Ribas, I., Lammer, H., Grießmeier, J.-M., Leitner, M., Selsis, F., Eiroa, C., Hanslmeier, A., Biernat, H. K., Farrugia, C. J., and Rucker, H. O. (2007). Coronal Mass Ejection (CME) activity of low mass M stars as an important factor for the habitability of terrestrial exoplanets. I. CME impact on expected magnetospheres of earth-like exoplanets in close-in habitable zones, *Astrobiology* **7**, pp. 167–184, doi:10.1089/ast.2006.0127.

Kislyakova, K. G., Johnstone, C. P., Odert, P., Erkaev, N. V., Lammer, H., Lüftinger, T., Holmström, M., Khodachenko, M. L., and Güdel, M. (2014). Stellar wind interaction and pick-up ion escape of the Kepler-11 "super-Earths", *A&A* **562**, A116, doi:10. 1051/0004-6361/201322933, arXiv:1312.4721 [astro-ph.EP].

Kislyakova, K. G., Lammer, H., Holmström, M., Panchenko, M., Odert, P., Erkaev, N. V., Leitzinger, M., Khodachenko, M. L., Kulikov, Y. N., Güdel, M., and Hanslmeier, A. (2013). XUV-exposed, non-hydrostatic hydrogen-rich upper atmospheres of terrestrial planets. Part II: Hydrogen coronae and ion escape, *Astrobiology* **13**, pp. 1030–1048, doi:10.1089/ast.2012.0958, arXiv:1212.4710 [astro-ph.EP].

Kley, W. and Nelson, R. P. (2010). Early evolution of planets in binaries: Planet-disk interaction, in N. Haghighipour (ed.), *Planets in Binary Star Systems* (Springer Netherlands, Dordrecht), ISBN 978-90-481-8687-7, pp. 135–164, doi:10.1007/978-90-481-8687-7_6, http://dx.doi.org/10.1007/978-90-481-8687-7_6.

Kokubo, E. and Ida, S. (1998). Oligarchic growth of protoplanets, *Icarus* **131**, pp. 171–178, doi:10.1006/icar.1997.5840.

Kopparapu, R. K., Ramirez, R., Kasting, J. F., Eymet, V., Robinson, T. D., Mahadevan, S., Terrien, R. C., Domagal-Goldman, S., Meadows, V., and Deshpande, R. (2013). Habitable zones around main-sequence stars: New estimates, *ApJ* **765**, 131, doi: 10.1088/0004-637X/765/2/131, arXiv:1301.6674 [astro-ph.EP].

Kopparapu, R. K., Ramirez, R. M., Schottelkotte, J., Kasting, J. F., Domagal-Goldman, S., and Eymet, V. (2014). Habitable zones around main-sequence stars: Dependence on planetary mass, *ApJL* **787**, L29, doi:10.1088/2041-8205/787/2/L29, arXiv:1404.5292 [astro-ph.EP].

Kostov, V. B., McCullough, P. R., Carter, J. A., Deleuil, M., Díaz, R. F., Fabrycky, D. C., Hébrard, G., Hinse, T. C., Mazeh, T., Orosz, J. A., Tsvetanov, Z. I., and Welsh, W. F. (2014). Kepler-413b: A slightly misaligned, neptune-size transiting circumbinary planet, *ApJ* **784**, 14, doi:10.1088/0004-637X/784/1/14, arXiv:1401.7275 [astro-ph.EP].

Kostov, V. B., McCullough, P. R., Hinse, T. C., Tsvetanov, Z. I., Hébrard, G., Díaz, R. F., Deleuil, M., and Valenti, J. A. (2013). A gas giant circumbinary planet transiting the F Star primary of the eclipsing binary star KIC 4862625 and the independent discovery and characterization of the two transiting planets in the Kepler-47 system, *ApJ* **770**, 52, doi:10.1088/0004-637X/770/1/52, arXiv:1210.3850 [astro-ph.EP].

Kostov, V. B., Orosz, J. A., Welsh, W. F., Doyle, L. R., Fabrycky, D. C., Haghighipour, N., Quarles, B., Short, D. R., Cochran, W. D., Endl, M., Ford, E. B., Gregorio, J., Hinse, T. C., Isaacson, H., Jenkins, J. M., Jensen, E. L. N., Kane, S., Kull, I., Latham, D. W., Lissauer, J. J., Marcy, G. W., Mazeh, T., Müller, T. W. A., Pepper, J., Quinn, S. N., Ragozzine, D., Shporer, A., Steffen, J. H., Torres, G., Windmiller, G., and Borucki, W. J. (2016). Kepler-1647b: The largest and longest-period Kepler transiting circumbinary planet, *ApJ* **827**, 86, doi:10.3847/0004-637X/827/1/86, arXiv:1512.00189 [astro-ph.EP].

Kozai, Y. (1962). Secular perturbations of asteroids with high inclination and eccentricity. *AJ* **67**, p. 579, doi:10.1086/108876.

Kraus, A. L., Andrews, S. M., Bowler, B. P., Herczeg, G., Ireland, M. J., Liu, M. C., Metchev, S., and Cruz, K. L. (2015). An ALMA disk mass for the candidate protoplanetary companion to FW Tau, *ApJL* **798**, L23, doi:10.1088/2041-8205/798/1/L23, arXiv:1412.2175 [astro-ph.EP].

Kraus, A. L., Ireland, M. J., Cieza, L. A., Hinkley, S., Dupuy, T. J., Bowler, B. P., and Liu, M. C. (2014). Three wide planetary-mass companions to FW Tau, ROXs 12, and ROXs 42B, *ApJ* **781**, 20, doi:10.1088/0004-637X/781/1/20, arXiv:1311.7664 [astro-ph.EP].

Kuzuhara, M., Tamura, M., Ishii, M., Kudo, T., Nishiyama, S., and Kandori, R. (2011). The widest-separation substellar companion candidate to a binary T Tauri Star, *AJ* **141**, 119, doi:10.1088/0004-6256/141/4/119.

Lagrange, A.-M., Beust, H., Udry, S., Chauvin, G., and Mayor, M. (2006). New constrains on Gliese 86 B. VLT near infrared coronographic imaging survey of planetary hosts, *A&A* **459**, pp. 955–963, doi:10.1051/0004-6361:20054710.

Lammer, H. (2007). M Star planet habitability, *Astrobiology* **7**, pp. 27–29, doi:10.1089/ast.2006.0123.

Lammer, H., Erkaev, N. V., Odert, P., Kislyakova, K. G., Leitzinger, M., and Khodachenko, M. L. (2013). Probing the blow-off criteria of hydrogen-rich 'super-Earths', *MNRAS* **430**, pp. 1247–1256, doi:10.1093/mnras/sts705, arXiv:1210.0793 [astro-ph.EP].

Lammer, H., Lichtenegger, H. I. M., Khodachenko, M. L., Kulikov, Y. N., and Griessmeier, J. (2011). The loss of nitrogen-rich atmospheres from Earth-like exoplanets within M-star habitable zones, in J. P. Beaulieu, S. Dieters, and G. Tinetti (eds.), *Molecules in the Atmospheres of Extrasolar Planets, Astronomical Society of the Pacific Conference Series*, Vol. 450, p. 139.

Lammer, H., Lichtenegger, H. I. M., Kulikov, Y. N., Grießmeier, J.-M., Terada, N., Erkaev, N. V., Biernat, H. K., Khodachenko, M. L., Ribas, I., Penz, T., and Selsis, F. (2007). Coronal Mass Ejection (CME) activity of low mass M Stars as an important factor for the habitability of terrestrial exoplanets. II. CME-induced ion pick up of Earth-like exoplanets in close-in habitable zones, *Astrobiology* **7**, pp. 185–207, doi:10.1089/ast.2006.0128.

Lammer, H., Selsis, F., Ribas, I., Guinan, E. F., Bauer, S. J., and Weiss, W. W. (2003). Atmospheric loss of exoplanets resulting from stellar x-ray and extreme-ultraviolet heating, *ApJL* **598**, pp. L121–L124, doi:10.1086/380815.

Lammer, H., Stökl, A., Erkaev, N. V., Dorfi, E. A., Odert, P., Güdel, M., Kulikov, Y. N., Kislyakova, K. G., and Leitzinger, M. (2014). Origin and loss of nebula-captured hydrogen envelopes from 'sub'- to 'super-Earths' in the habitable zone of Sun-like stars, *MNRAS* **439**, pp. 3225–3238, doi:10.1093/mnras/stu085, arXiv:1401.2765 [astro-ph.EP].

Larson, R. B. (1972). The collapse of a rotating cloud, *MNRAS* **156**, p. 437, doi:10.1093/mnras/156.4.437.

Laskar, J. (1988). Secular evolution of the solar system over 10 million years, *A&A* **198**, pp. 341–362.

Laskar, J. and Boué, G. (2010). Explicit expansion of the three-body disturbing function for arbitrary eccentricities and inclinations, *A&A* **522**, A60, doi:10.1051/0004-6361/201014496, arXiv:1008.2947 [astro-ph.IM].

Laskar, J. and Robutel, P. (1995). Stability of the planetary three-body problem. I. Expansion of the planetary Hamiltonian, *CMDA* **62**, pp. 193–217, doi:10.1007/BF00692088.

Lecar, M., Podolak, M., Sasselov, D., and Chiang, E. (2006). On the location of the snow line in a protoplanetary disk, *ApJ* **640**, pp. 1115–1118, doi:10.1086/500287, astro-ph/0602217.

Leconte, J., Forget, F., Charnay, B., Wordsworth, R., Selsis, F., Millour, E., and Spiga, A. (2013). 3D climate modeling of close-in land planets: Circulation patterns, climate moist bistability, and habitability, *A&A* **554**, A69, doi:10.1051/0004-6361/201321042, arXiv:1303.7079 [astro-ph.EP].

Lee, J. W., Hinse, T. C., Youn, J.-H., and Han, W. (2014). The pulsating sdB+M eclipsing system NY Virginis and its circumbinary planets, *MNRAS* **445**, pp. 2331–2339, doi:10.1093/mnras/stu1937, arXiv:1409.4907 [astro-ph.SR].

Leinhardt, Z. M. and Stewart, S. T. (2012). Collisions between gravity-dominated bodies. I. Outcome regimes and scaling laws, *ApJ* **745**, 79, doi:10.1088/0004-637X/745/1/79, arXiv:1106.6084 [astro-ph.EP].

Lemaître, A. (2010). Resonances: Models and captures, in J. Souchay and R. Dvorak (eds.), *Lecture Notes in Physics, Berlin Springer Verlag, Lecture Notes in Physics, Berlin Springer Verlag*, Vol. 790, pp. 1–62, doi:10.1007/978-3-642-04458-8_1.

Levison, H. F. and Agnor, C. (2003). The role of giant planets in terrestrial planet formation, *AJ* **125**, pp. 2692–2713, doi:10.1086/374625.

Lichtenegger, H. I. M., Kislyakova, K. G., Odert, P., Erkaev, N. V., Lammer, H., Gröller, H., Johnstone, C. P., Elkins-Tanton, L., Tu, L., Güdel, M., and Holmström, M. (2016). Solar XUV and ENA-driven water loss from early Venus' steam atmosphere, *Journal of Geophysical Research (Space Physics)* **121**, pp. 4718–4732, doi:10.1002/2015JA022226.

Lichtenegger, H. I. M., Lammer, H., Grießmeier, J.-M., Kulikov, Y. N., von Paris, P., Hausleitner, W., Krauss, S., and Rauer, H. (2010). Aeronomical evidence for higher CO_2 levels during Earth's Hadean epoch, *Icarus* **210**, pp. 1–7, doi:10.1016/j.icarus.2010.06.042.

Lissauer, J. J., Quintana, E. V., Chambers, J. E., Duncan, M. J., and Adams, F. C. (2004). Terrestrial planet formation in binary star systems, in G. Garcia-Segura, G. Tenorio-Tagle, J. Franco, and H. W. Yorke (eds.), *Revista Mexicana de Astronomia y Astrofisica Conference Series, Revista Mexicana de Astronomia y Astrofisica Conference Series*, Vol. 22, pp. 99–103.

Lo Curto, G., Mayor, M., Clausen, J. V., Benz, W., Bouchy, F., Lovis, C., Moutou, C., Naef, D., Pepe, F., Queloz, D., Santos, N. C., Sivan, J.-P., Udry, S., Bonfils, X., Delfosse, X., Mordasini, C., Fouqué, P., Olsen, E. H., and Pritchard, J. D. (2006). The HARPS search for southern extra-solar planets. VII. A very hot Jupiter orbiting HD 212301, *A&A* **451**, pp. 345–350, doi:10.1051/0004-6361:20054083.

Ludwig, W., Eggl, S., Neubauer, D., Leitner, J., Firneis, M. G., and Hitzenberger, R. (2016). Effective stellar flux calculations for limits of life-supporting zones of exoplanets, *MNRAS* **458**, pp. 3752–3759, doi:10.1093/mnras/stw509.

Maindl, T. I., Schäfer, C., Speith, R., Süli, Á., Forgács-Dajka, E., and Dvorak, R. (2013). SPH-based simulation of multi-material asteroid collisions, *Astronomische Nachrichten* **334**, 9, pp. 996–999.

Marchal, C. (1990). The three-body problem, *Studies in Aeronautics*, 4 *(Elsevier, Amsterdam, NED)*.

Marcy, G. W., Butler, R. P., and Vogt, S. S. (2000). Sub-Saturn planetary candidates of HD 16141 and HD 46375, *ApJL* **536**, pp. L43–L46, doi:10.1086/312723, astro-ph/0004326.

Mardling, R. A. (2007). Long-term tidal evolution of short-period planets with companions, *MNRAS* **382**, pp. 1768–1790, doi:10.1111/j.1365-2966.2007.12500.x, arXiv:0706.0224.

Mardling, R. A. (2013). New developments for modern celestial mechanics — I. General coplanar three-body systems. Application to exoplanets, *MNRAS* **435**, pp. 2187–2226, doi:10.1093/mnras/stt1438, http://arxiv.org/abs/1308.0607arXiv:1308.0607 [astro-ph.EP].

Mardling, R. A. and Aarseth, S. J. (2001). Tidal interactions in star cluster simulations, *MNRAS* **321**, pp. 398–420, doi:10.1046/j.1365-8711.2001.03974.x.

Martin, R. G. and Livio, M. (2012). On the evolution of the snow line in protoplanetary discs, *MNRAS* **425**, pp. L6–L9, arXiv:1207.4284 [astro-ph.EP].

Martin, R. G. and Livio, M. (2013). On the evolution of the snow line in protoplanetary discs — II. Analytic approximations, *MNRAS* **434**, pp. 633–638, arXiv:1306.5243 [astro-ph.EP].

Martioli, E., Colón, K. D., Angerhausen, D., Stassun, K. G., Rodriguez, J. E., Zhou, G., Gaudi, B. S., Pepper, J., Beatty, T. G., Tata, R., James, D. J., Eastman, J. D., Wilson, P. A., Bayliss, D., and Stevens, D. J. (2018). A survey of eight hot Jupiters in secondary eclipse using WIRCam at CFHT, *MNRAS* **474**, pp. 4264–4277, doi:10.1093/mnras/stx3009, arXiv:1711.07294 [astro-ph.EP].

Marzari, F. and Scholl, H. (2000). Planetesimal accretion in binary star systems, *ApJ* **543**, pp. 328–339, doi:10.1086/317091.

Mason, P. A., Zuluaga, J. I., Clark, J. M., and Cuartas-Restrepo, P. A. (2013). Rotational synchronization may enhance habitability for circumbinary planets: Kepler binary case studies, *ApJL* **774**, L26, doi:10.1088/2041-8205/774/2/L26, arXiv:1307.4624 [astro-ph.EP].

Mason, P. A., Zuluaga, J. I., Cuartas-Restrepo, P. A., and Clark, J. M. (2015). Circumbinary habitability niches, *International Journal of Astrobiology* **14**, pp. 391–400, doi:10.1017/S1473550414000342, arXiv:1408.5163 [astro-ph.EP].

Maxted, P. F. L., Anderson, D. R., Collier Cameron, A., Doyle, A. P., Fumel, A., Gillon, M., Hellier, C., Jehin, E., Lendl, M., Pepe, F., Pollacco, D. L., Queloz, D., Ségransan, D., Smalley, B., Southworth, K., Smith, A. M. S., Triaud, A. H. M. J., Udry, S., and West, R. G. (2013). WASP-77 Ab: A transiting hot Jupiter planet in a wide binary system, *PASP* **125**, p. 48, doi:10.1086/669231, arXiv:1211.6033 [astro-ph.EP].

Mayor, M., Udry, S., Naef, D., Pepe, F., Queloz, D., Santos, N. C., and Burnet, M. (2004). The CORALIE survey for southern extra-solar planets. XII. Orbital solutions for 16 extra-solar planets discovered with CORALIE, *A&A* **415**, pp. 391–402, doi:10.1051/0004-6361:20034250, astro-ph/0310316.

Mills, S. M. and Mazeh, T. (2017). The planetary mass-radius relation and its dependence on orbital period as measured by transit timing variations and radial velocities, *ApJL* **839**, L8, doi:10.3847/2041-8213/aa67eb, arXiv:1703.07790 [astro-ph.EP].

Morbidelli, A. (2002). *Modern Celestial Mechanics: Aspects of Solar System Dynamics* (Taylor & Francis, London), doi:ISBN0415279399.

Morbidelli, A., Chambers, J., Lunine, J. I., Petit, J. M., Robert, F., Valsecchi, G. B., and Cyr, K. E. (2000). Source regions and time scales for the delivery of water to Earth, *Meteoritics and Planetary Science* **35**, pp. 1309–1320.

Moriwaki, K. and Nakagawa, Y. (2004). A planetesimal accretion zone in a circumbinary disk, *ApJ* **609**, pp. 1065–1070, doi:10.1086/421342.

Mugrauer, M. and Neuhäuser, R. (2005). Gl86B: A white dwarf orbits an exoplanet host star, *MNRAS* **361**, pp. L15–L19, doi:10.1111/j.1745-3933.2005.00055.x, astro-ph/0506311.

Mugrauer, M. and Neuhäuser, R. (2009). The multiplicity of exoplanet host stars. New low-mass stellar companions of the exoplanet host stars HD 125612 and HD 212301, *A&A* **494**, pp. 373–378, doi:10.1051/0004-6361:200810639, arXiv:0812.2561.

Mugrauer, M., Neuhäuser, R., and Mazeh, T. (2007a). The multiplicity of exoplanet host stars. Spectroscopic confirmation of the companions GJ 3021 B and HD 27442 B, one new planet host triple-star system, and global statistics, *A&A* **469**, pp. 755–770, doi:10.1051/0004-6361:20065883, astro-ph/0703795.

Mugrauer, M., Neuhäuser, R., Mazeh, T., Guenther, E., Fernández, M., and Broeg, C. (2006). A search for wide visual companions of exoplanet host stars: The Calar Alto Survey, *Astronomische Nachrichten* **327**, p. 321, doi:10.1002/asna.200510528, astro-ph/0602067.

Mugrauer, M., Neuhäuser, R., Seifahrt, A., Mazeh, T., and Guenther, E. (2005). Four new wide binaries among exoplanet host stars, *A&A* **440**, pp. 1051–1060, doi:10.1051/0004-6361:20042297, astro-ph/0507101.

Mugrauer, M., Seifahrt, A., and Neuhäuser, R. (2007b). The multiplicity of planet host stars — New low-mass companions to planet host stars, *MNRAS* **378**, pp. 1328–1334, doi:10.1111/j.1365-2966.2007.11858.x, arXiv:0704.1767.

Müller, T. W. A. and Haghighipour, N. (2014). Calculating the habitable zones of multiple star systems with a new interactive web site, *ApJ* **782**, 26, doi:10.1088/0004-637X/782/1/26, arXiv:1401.0601 [astro-ph.EP].

Müller, T. W. A. and Kley, W. (2012). Circumstellar disks in binary star systems. Models for γ Cephei and α Centauri, *A&A* **539**, A18, doi:10.1051/0004-6361/201118202, arXiv:1112.1845 [astro-ph.EP].

Murray, C. D. and Dermott, S. F. (1999). *Solar System Dynamics* (Cambridge University Press).

Naef, D., Mayor, M., Pepe, F., Queloz, D., Santos, N. C., Udry, S., and Burnet, M. (2001). The CORALIE survey for southern extrasolar planets. V. 3 new extrasolar planets, *A&A* **375**, pp. 205–218, doi:10.1051/0004-6361:20010841, astro-ph/0106255.

Nesvorný, D. and Morbidelli, A. (1998). Three-body mean motion resonances and the chaotic structure of the asteroid belt, *AJ* **116**, pp. 3029–3037, doi:10.1086/300632.

Neuhäuser, R., Mugrauer, M., Fukagawa, M., Torres, G., and Schmidt, T. (2007). Direct detection of exoplanet host star companion γ Cep B and revised masses for both stars and the sub-stellar object, *A&A* **462**, pp. 777–780, doi:10.1051/0004-6361:20066581, arXiv:astro-ph/0611427.

O'Donovan, F. T., Charbonneau, D., Harrington, J., Madhusudhan, N., Seager, S., Deming, D., and Knutson, H. A. (2010). Detection of planetary emission from the exoplanet Tres-2 using Spitzer/IRAC, *ApJ* **710**, pp. 1551–1556, doi:10.1088/0004-637X/710/2/1551, arXiv:0909.3073 [astro-ph.EP].

O'Donovan, F. T., Charbonneau, D., Mandushev, G., Dunham, E. W., Latham, D. W., Torres, G., Sozzetti, A., Brown, T. M., Trauger, J. T., Belmonte, J. A., Rabus, M., Almenara, J. M., Alonso, R., Deeg, H. J., Esquerdo, G. A., Falco, E. E., Hillenbrand, L. A., Roussanova, A., Stefanik, R. P., and Winn, J. N. (2006). TrES-2: The first transiting planet in the Kepler field, *ApJL* **651**, pp. L61–L64, doi:10.1086/509123, astro-ph/0609335.

Orosz, J. A., Welsh, W. F., Carter, J. A., Brugamyer, E., Buchhave, L. A., Cochran, W. D., Endl, M., Ford, E. B., MacQueen, P., Short, D. R., Torres, G., Windmiller, G., Agol, E., Barclay, T., Caldwell, D. A., Clarke, B. D., Doyle, L. R., Fabrycky, D. C., Geary, J. C., Haghighipour, N., Holman, M. J., Ibrahim, K. A., Jenkins, J. M., Kinemuchi, K., Li, J., Lissauer, J. J., Prša, A., Ragozzine, D., Shporer, A., Still, M., and Wade, R. A. (2012a). The Neptune-sized circumbinary planet Kepler-38b, *ApJ* **758**, 87, doi:10.1088/0004-637X/758/2/87, arXiv:1208.3712 [astro-ph.SR].

Orosz, J. A., Welsh, W. F., Carter, J. A., Fabrycky, D. C., Cochran, W. D., Endl, M., Ford, E. B., Haghighipour, N., MacQueen, P. J., Mazeh, T., Sanchis-Ojeda, R., Short, D. R., Torres, G., Agol, E., Buchhave, L. A., Doyle, L. R., Isaacson, H., Lissauer, J. J., Marcy, G. W., Shporer, A., Windmiller, G., Barclay, T., Boss, A. P., Clarke, B. D., Fortney, J., Geary, J. C., Holman, M. J., Huber, D., Jenkins, J. M., Kinemuchi, K., Kruse, E., Ragozzine, D., Sasselov, D., Still, M., Tenenbaum, P., Uddin, K., Winn, J. N., Koch, D. G., and Borucki, W. J. (2012b). Kepler-47: A transiting circumbinary multiplanet system, *Science* **337**, p. 1511, doi:10.1126/science.1228380, arXiv:1208.5489 [astro-ph.SR].

Pichardo, B., Sparke, L. S., and Aguilar, L. A. (2005). Circumstellar and circumbinary discs in eccentric stellar binaries, *MNRAS* **359**, pp. 521–530, doi:10.1111/j.1365-2966.2005.08905.x, arXiv:astro-ph/0501244.

Pilat-Lohinger, E. (2005). Planetary motion in double stars: The influence of the secondary, in Z. Knežević and A. Milani (eds.), *IAU Colloq. 197: Dynamics of Populations of Planetary Systems*, pp. 71–76, doi:10.1017/S1743921304008518.

Pilat-Lohinger, E. (2012). Dynamical stability and habitability of extra-solar planets, in M. T. Richards and I. Hubeny (eds.), *From Interacting Binaries to Exoplanets: Essential Modeling Tools*, *IAU Symposium*, Vol. 282, pp. 539–544, doi:10.1017/S174392131102833X.

Pilat-Lohinger, E. (2015). The role of dynamics on the habitability of an Earth-like planet, *International Journal of Astrobiology* **14**, pp. 145–152, doi:10.1017/S1473550414000469, arXiv:1505.07039 [astro-ph.EP].

Pilat-Lohinger, E., Bazsó, Á., and Funk, B. (2016). A quick method to identify secular resonances in multi-planet systems with a binary companion, *AJ* **152**, 139, doi:10.3847/0004-6256/152/5/139.

Pilat-Lohinger, E. and Dvorak, R. (2002). Stability of S-type Orbits in binaries, *Celestial Mechanics and Dynamical Astronomy* **82**, pp. 143–153, doi:10.1023/A: 1014586308539.

Pilat-Lohinger, E. and Funk, B. (2010). Dynamical stability of extra-solar planets, in J. Souchay and R. Dvorak (eds.), *Lecture Notes in Physics, Berlin Springer Verlag, Lecture Notes in Physics, Berlin Springer Verlag*, Vol. 790, pp. 481–510, doi:10. 1007/978-3-642-04458-8_10.

Pilat-Lohinger, E., Funk, B., and Dvorak, R. (2003). Stability limits in double stars. A study of inclined planetary orbits, *A&A* **400**, pp. 1085–1094, doi:10.1051/0004-6361: 20021811.

Pilat-Lohinger, E., Robutel, P., Süli, Á., and Freistetter, F. (2008a). On the stability of Earth-like planets in multi-planet systems, *CMDA* **102**, pp. 83–95, doi:10.1007/ s10569-008-9159-0.

Pilat-Lohinger, E., Süli, Á., Robutel, P., and Freistetter, F. (2008b). The influence of giant planets near a mean motion resonance on earth-like planets in the habitable zone of sun-like stars, *ApJ* **681**, pp. 1639–1645, doi:10.1086/587501.

Popp, M. and Eggl, S. (2017). Climate variations on Earth-like circumbinary planets, *Nature Communications* **8**, 14957, doi:10.1038/ncomms14957.

Popp, M., Schmidt, H., and Marotzke, J. (2016). Transition to a moist greenhouse with CO_2 and solar forcing, *Nature Communications* **7**, 10627, doi:10.1038/ncomms10627, arXiv:1610.02283 [astro-ph.EP].

Qian, S.-B., Liao, W.-P., Zhu, L.-Y., and Dai, Z.-B. (2010). Detection of a giant extrasolar planet orbiting the eclipsing polar DP Leo, *ApJL* **708**, pp. L66–L68, doi:10.1088/ 2041-8205/708/1/L66.

Qian, S.-B., Liu, L., Liao, W.-P., Li, L.-J., Zhu, L.-Y., Dai, Z.-B., He, J.-J., Zhao, E.-G., Zhang, J., and Li, K. (2011). Detection of a planetary system orbiting the eclipsing polar HU Aqr, *MNRAS* **414**, pp. L16–L20, doi:10.1111/j.1745-3933.2011.01045.x, arXiv:1103.2005 [astro-ph.SR].

Qian, S.-B., Liu, L., Zhu, L.-Y., Dai, Z.-B., Fernández Lajús, E., and Baume, G. L. (2012a). A circumbinary planet in orbit around the short-period white dwarf eclipsing binary RR Cae, *MNRAS* **422**, pp. 24–27, doi:10.1111/j.1745-3933.2012.01228.x, arXiv:1201.4205 [astro-ph.SR].

Qian, S.-B., Zhu, L.-Y., Dai, Z.-B., Fernández-Lajús, E., Xiang, F.-Y., and He, J.-J. (2012b). Circumbinary planets orbiting the rapidly pulsating subdwarf B-type binary NY Vir, *ApJL* **745**, L23, doi:10.1088/2041-8205/745/2/L23, arXiv:1112.4269 [astro-ph.SR].

Queloz, D., Mayor, M., Weber, L., Blécha, A., Burnet, M., Confino, B., Naef, D., Pepe, F., Santos, N., and Udry, S. (2000). The CORALIE survey for southern extra-solar planets. I. A planet orbiting the star Gliese 86, *A&A* **354**, pp. 99–102.

Quintana, E. V., Adams, F. C., Lissauer, J. J., and Chambers, J. E. (2007). Terrestrial planet formation around individual stars within binary star systems, *ApJ* **660**, pp. 807–822, doi:10.1086/512542, astro-ph/0701266.

Rabl, G. and Dvorak, R. (1988). Satellite-type planetary orbits in double stars — A numerical approach, *A&A* **191**, pp. 385–391.

Rabus, M., Deeg, H. J., Alonso, R., Belmonte, J. A., and Almenara, J. M. (2009). Transit timing analysis of the exoplanets TrES-1 and TrES-2, *A&A* **508**, pp. 1011–1020, doi:10.1051/0004-6361/200912252, arXiv:0909.1564 [astro-ph.EP].

Raetz, S., Maciejewski, G., Ginski, C., Mugrauer, M., Berndt, A., Eisenbeiss, T., Adam, C., Raetz, M., Roell, T., Seeliger, M., Marka, C., Vaňko, M., Bukowiecki, Ł., Errmann, R., Kitze, M., Ohlert, J., Pribulla, T., Schmidt, J. G., Sebastian, D., Puchalski, D., Tetzlaff, N., Hohle, M. M., Schmidt, T. O. B., and Neuhäuser, R. (2014). Transit timing of TrES-2: A combined analysis of ground- and space-based photometry, *MNRAS* **444**, pp. 1351–1368, doi:10.1093/mnras/stu1505, arXiv:1408.7022 [astro-ph.SR].

Rafikov, R. R. (2013). Building Tatooine: Suppression of the direct secular excitation in Kepler circumbinary planet formation, *ApJL* **764**, L16, doi:10.1088/2041-8205/764/1/L16, arXiv:1212.2217 [astro-ph.EP].

Raghavan, D., Henry, T. J., Mason, B. D., Subasavage, J. P., Jao, W.-C., Beaulieu, T. D., and Hambly, N. C. (2006). Two suns in the sky: Stellar multiplicity in exoplanet systems, *ApJ* **646**, pp. 523–542, doi:10.1086/504823, astro-ph/0603836.

Raghavan, D., McAlister, H. A., Henry, T. J., Latham, D. W., Marcy, G. W., Mason, B. D., Gies, D. R., White, R. J., and ten Brummelaar, T. A. (2010). A survey of stellar families: Multiplicity of solar-type stars, *ApJS* **190**, pp. 1–42, doi:10.1088/0067-0049/190/1/1, arXiv:1007.0414 [astro-ph.SR].

Ramirez, R. M. and Kaltenegger, L. (2016). Habitable zones of post-main sequence stars, *ApJ* **823**, 6, doi:10.3847/0004-637X/823/1/6, arXiv:1605.04924 [astro-ph.EP].

Rasool, S. I. and de Bergh, C. (1970). The runaway greenhouse and the accumulation of CO_2 in the Venus atmosphere, *Nature* **226**, pp. 1037–1039, doi:10.1038/2261037a0.

Reegen, P. (2007). SigSpec. I. Frequency- and phase-resolved significance in Fourier space, *A&A* **467**, pp. 1353–1371, doi:10.1051/0004-6361:20066597, physics/0703160.

Roberts, L. C., Jr., Turner, N. H., ten Brummelaar, T. A., Mason, B. D., and Hartkopf, W. I. (2011). Know the star, know the planet. I. Adaptive optics of exoplanet host stars, *AJ* **142**, 175, doi:10.1088/0004-6256/142/5/175, arXiv:1109.4320 [astro-ph.SR].

Roell, T., Neuhäuser, R., Seifahrt, A., and Mugrauer, M. (2012). Extrasolar planets in stellar multiple systems, *A&A* **542**, A92, doi:10.1051/0004-6361/201118051, arXiv:1204.4833 [astro-ph.SR].

Roy, A. E. (1988). *Orbital Motion*, 3rd revised and enlarged edition edn. (Adam Hilger, Bristol, England).

Salz, M., Schneider, P. C., Czesla, S., and Schmitt, J. H. M. M. (2015). High-energy irradiation and mass loss rates of hot Jupiters in the solar neighborhood, *A&A* **576**, A42, doi:10.1051/0004-6361/201425243, arXiv:1502.00576 [astro-ph.EP].

Santerne, A., Hébrard, G., Deleuil, M., Havel, M., Correia, A. C. M., Almenara, J.-M., Alonso, R., Arnold, L., Barros, S. C. C., Behrend, R., Bernasconi, L., Boisse, I., Bonomo, A. S., Bouchy, F., Bruno, G., Damiani, C., Díaz, R. F., Gravallon, D., Guillot, T., Labrevoir, O., Montagnier, G., Moutou, C., Rinner, C., Santos, N. C., Abe, L., Audejean, M., Bendjoya, P., Gillier, C., Gregorio, J., Martinez, P.,

Michelet, J., Montaigut, R., Poncy, R., Rivet, J.-P., Rousseau, G., Roy, R., Suarez, O., Vanhuysse, M., and Verilhac, D. (2014). SOPHIE velocimetry of Kepler transit candidates. XII. KOI-1257 b: a highly eccentric three-month period transiting exoplanet, *A&A* **571**, A37, doi:10.1051/0004-6361/201424158, arXiv:1406.6172 [astro-ph.EP].

Santos, N. C., Mayor, M., Naef, D., Pepe, F., Queloz, D., Udry, S., and Blecha, A. (2000). The CORALIE survey for Southern extra-solar planets. IV. Intrinsic stellar limitations to planet searches with radial-velocity techniques, *A&A* **361**, pp. 265–272.

Savonije, G. J., Papaloizou, J. C. B., and Lin, D. N. C. (1994). On tidally induced shocks in accretion discs in close binary systems, *MNRAS* **268**, p. 13.

Schäfer, C. (2005). *Application of Smooth Particle Hydrodynamics to Selected Aspects of Planet Formation*, Dissertation, Eberhard-Karls-Universität Tübingen.

Schäfer, C., Riecker, S., Maindl, T. I., Speith, R., Scherrer, S., and Kley, W. (2016). A smooth particle hydrodynamics code to model collisions between solid, self-gravitating objects, *A&A* **590**, A19, arXiv:1604.03290 [astro-ph.EP].

Schneider, J., Dedieu, C., Le Sidaner, P., Savalle, R., and Zolotukhin, I. (2011). Defining and cataloging exoplanets: the exoplanet.eu database, *A&A* **532**, A79, doi:10.1051/0004-6361/201116713, arXiv:1106.0586 [astro-ph.EP].

Schwarz, R., Funk, B., and Bazsó, Á. (2015). On the possibility of habitable trojan planets in binary star systems, *Origins of Life and Evolution of the Biosphere* **45**, pp. 469–477, doi:10.1007/s11084-015-9449-y.

Schwarz, R., Funk, B., Zechner, R., and Bazsó, Á. (2016). New prospects for observing and cataloguing exoplanets in well-detached binaries, *MNRAS* **460**, pp. 3598–3609, doi:10.1093/mnras/stw1218, arXiv:1608.00764 [astro-ph.EP].

Sekiya, M., Miyama, S. M., and Hayashi, C. (1988). Chapter 23. Gas capture by Proto-Jupiter and Proto-Saturn, *Progress of Theoretical Physics Supplement* **96**, pp. 274–280, doi:10.1143/PTPS.96.274.

Selsis, F., Kasting, J. F., Levrard, B., Paillet, J., Ribas, I., and Delfosse, X. (2007). Habitable planets around the star Gliese 581? *A&A* **476**, pp. 1373–1387, doi:10.1051/0004-6361:20078091, arXiv:0710.5294.

Sigurdsson, S., Richer, H. B., Hansen, B. M., Stairs, I. H., and Thorsett, S. E. (2003). A young white dwarf companion to Pulsar B1620-26: Evidence for early planet formation, *Science* **301**, pp. 193–196, doi:10.1126/science.1086326, astro-ph/0307339.

Silsbee, K. and Rafikov, R. R. (2015). Planet formation in binaries: Dynamics of planetesimals perturbed by the eccentric protoplanetary disk and the secondary, *ApJ* **798**, 71, doi:10.1088/0004-637X/798/2/71, arXiv:1309.3290 [astro-ph.EP].

Sirothia, S. K., Lecavelier des Etangs, A., Gopal-Krishna, Kantharia, N. G., and Ishwar-Chandra, C. H. (2014). Search for 150 MHz radio emission from extrasolar planets in the TIFR GMRT Sky Survey, *A&A* **562**, A108, doi:10.1051/0004-6361/201321571.

Sitarski, G. (1968). Approaches of the parabolic comets to the outer planets, *Acta Astronom.* **18**, p. 171.

Slawson, R. W., Prša, A., Welsh, W. F., Orosz, J. A., Rucker, M., Batalha, N., Doyle, L. R., Engle, S. G., Conroy, K., Coughlin, J., Gregg, T. A., Fetherolf, T., Short, D. R., Windmiller, G., Fabrycky, D. C., Howell, S. B., Jenkins, J. M., Uddin, K., Mullally, F., Seader, S. E., Thompson, S. E., Sanderfer, D. T., Borucki, W., and Koch, D. (2011). Kepler eclipsing binary stars. II. 2165 eclipsing binaries in the second data release, *AJ* **142**, 160, doi:10.1088/0004-6256/142/5/160, arXiv:1103.1659 [astro-ph.SR].

Southworth, J. (2012). Homogeneous studies of transiting extrasolar planets — V. New results for 38 planets, *MNRAS* **426**, pp. 1291–1323, doi:10.1111/j.1365-2966.2012. 21756.x, arXiv:1207.5796 [astro-ph.EP].

Southworth, J., Mancini, L., Novati, S. C., Dominik, M., Glitrup, M., Hinse, T. C., Jørgensen, U. G., Mathiasen, M., Ricci, D., Maier, G., Zimmer, F., Bozza, V., Browne, P., Bruni, I., Burgdorf, M., Dall'Ora, M., Finet, F., Harpsøe, K., Hundertmark, M., Liebig, C., Rahvar, S., Scarpetta, G., Skottfelt, J., Smalley, B., Snodgrass, C., and Surdej, J. (2010). High-precision photometry by telescope defocusing — III. The transiting planetary system WASP-2, *MNRAS* **408**, pp. 1680–1688, doi:10.1111/j.1365-2966. 2010.17238.x, arXiv:1006.4464 [astro-ph.EP].

Spiegel, D. S., Menou, K., and Scharf, C. A. (2008). Habitable climates, *ApJ* **681**, pp. 1609–1623, doi:10.1086/588089, arXiv:0711.4856.

Spiegel, D. S., Raymond, S., Dressing, C. D., Scharf, C. A., Mitchell, J. L., and Menou, K. (2010). General Milankovitch cycles, in V. Coudé Du Foresto, D. M. Gelino, and I. Ribas (eds.), *Pathways Towards Habitable Planets, Astronomical Society of the Pacific Conference Series*, Vol. 430, p. 109.

Stewart, S. T. and Leinhardt, Z. M. (2009). Velocity-dependent catastrophic disruption criteria for planetesimals, *ApJL* **691**, pp. L133–L137, doi:10.1088/0004-637X/691/ 2/L133.

Stewart, S. T. and Leinhardt, Z. M. (2012). Collisions between gravity-dominated bodies. II. The diversity of impact outcomes during the end stage of planet formation, *ApJ* **751**, 32, doi:10.1088/0004-637X/751/1/32, arXiv:1109.4588 [astro-ph.EP].

Stökl, A., Dorfi, E. A., Johnstone, C. P., and Lammer, H. (2016). Dynamical accretion of primordial atmospheres around planets with masses between 0.1 and 5 $M_\oplus$ in the habitable zone, *ApJ* **825**, 86, doi:10.3847/0004-637X/825/2/86.

Szebehely, V. (1980). Stability of planetary orbits in binary systems, *Celestial Mechanics* **22**, pp. 7–12, doi:10.1007/BF01228750.

Szebehely, V. and McKenzie, R. (1981). Stability of outer planetary systems, *Celestial Mechanics* **23**, pp. 3–7, doi:10.1007/BF01228541.

Thébault, P. (2011). Against all odds? Forming the planet of the HD 196885 binary, *CMDA* **111**, pp. 29–49, doi:10.1007/s10569-011-9346-2, arXiv:1103.3900 [astro-ph.EP].

Thébault, P., Marzari, F., and Scholl, H. (2006). Relative velocities among accreting planetesimals in binary systems: The circumprimary case, *Icarus* **183**, pp. 193–206, doi:10.1016/j.icarus.2006.01.022, arXiv:astro-ph/0602046.

Thébault, P., Marzari, F., Scholl, H., Turrini, D., and Barbieri, M. (2004). Planetary formation in the γ Cephei system, *A&A* **427**, pp. 1097–1104, doi:10.1051/0004-6361: 20040514, arXiv:astro-ph/0408153.

Thévenin, F., Provost, J., Morel, P., Berthomieu, G., Bouchy, F., and Carrier, F. (2002). Asteroseismology and calibration of alpha Cen binary system, *A&A* **392**, pp. L9–L12, doi:10.1051/0004-6361:20021074, astro-ph/0206283.

Thorsett, S. E., Arzoumanian, Z., and Taylor, J. H. (1993). PSR B1620-26 — A binary radio pulsar with a planetary companion? *ApJL* **412**, pp. L33–L36, doi:10.1086/186933.

Tian, F., Kasting, J. F., Liu, H.-L., and Roble, R. G. (2008). Hydrodynamic planetary thermosphere model: 1. Response of the Earth's thermosphere to extreme solar EUV conditions and the significance of adiabatic cooling, *Journal of Geophysical Research (Planets)* **113**, E05008, doi:10.1029/2007JE002946.

Tian, F., Toon, O. B., Pavlov, A. A., and De Sterck, H. (2005). Transonic hydrodynamic escape of hydrogen from extrasolar planetary atmospheres, *ApJ* **621**, pp. 1049–1060, doi:10.1086/427204.

Tokovinin, A. (2014). From binaries to multiples. II. Hierarchical multiplicity of F and G dwarfs, *AJ* **147**, 87, doi:10.1088/0004-6256/147/4/87, arXiv:1401.6827 [astro-ph.SR].

Trifonov, T., Kürster, M., Zechmeister, M., Tal-Or, L., Caballero, J. A., Quirrenbach, A., Amado, P. J., Ribas, I., Reiners, A., Reffert, S., Dreizler, S., Hatzes, A. P., Kaminski, A., Launhardt, R., Henning, T., Montes, D., Béjar, V. J. S., Mundt, R., Pavlov, A., Schmitt, J. H. M. M., Seifert, W., Morales, J. C., Nowak, G., Jeffers, S. V., Rodríguez-López, C., del Burgo, C., Anglada-Escudé, G., López-Santiago, J., Mathar, R. J., Ammler-von Eiff, M., Guenther, E. W., Barrado, D., González Hernández, J. I., Mancini, L., Stürmer, J., Abril, M., Aceituno, J., Alonso-Floriano, F. J., Antona, R., Anwand-Heerwart, H., Arroyo-Torres, B., Azzaro, M., Baroch, D., Bauer, F. F., Becerril, S., Benítez, D., Berdiñas, Z. M., Bergond, G., Blümcke, M., Brinkmöller, M., Cano, J., Cárdenas Vázquez, M. C., Casal, E., Cifuentes, C., Claret, A., Colomé, J., Cortés-Contreras, M., Czesla, S., Díez-Alonso, E., Feiz, C., Fernández, M., Ferro, I. M., Fuhrmeister, B., Galadí-Enríquez, D., Garcia-Piquer, A., García Vargas, M. L., Gesa, L., Gómez Galera, V., González-Peinado, R., Grözinger, U., Grohnert, S., Guàrdia, J., Guijarro, A., de Guindos, E., Gutiérrez-Soto, J., Hagen, H.-J., Hauschildt, P. H., Hedrosa, R. P., Helmling, J., Hermelo, I., Hernández Arabí, R., Hernández Castaño, L., Hernández Hernando, F., Herrero, E., Huber, A., Huke, P., Johnson, E., de Juan, E., Kim, M., Klein, R., Klüter, J., Klutsch, A., Lafarga, M., Lampón, M., Lara, L. M., Laun, W., Lemke, U., Lenzen, R., López del Fresno, M., López-González, J., López-Puertas, M., López Salas, J. F., Luque, R., Magán Madinabeitia, H., Mall, U., Mandel, H., Marfil, E., Marín Molina, J. A., Maroto Fernández, D., Martín, E. L., Martín-Ruiz, S., Marvin, C. J., Mirabet, E., Moya, A., Moreno-Raya, M. E., Nagel, E., Naranjo, V., Nortmann, L., Ofir, A., Oreiro, R., Pallé, E., Panduro, J., Pascual, J., Passegger, V. M., Pedraz, S., Pérez-Calpena, A., Pérez Medialdea, D., Perger, M., Perryman, M. A. C., Pluto, M., Rabaza, O., Ramón, A., Rebolo, R., Redondo, P., Reinhardt, S., Rhode, P., Rix, H.-W., Rodler, F., Rodríguez, E., Rodríguez Trinidad, A., Rohloff, R.-R., Rosich, A., Sadegi, S., Sánchez-Blanco, E., Sánchez Carrasco, M. A., Sánchez-López, A., Sanz-Forcada, J., Sarkis, P., Sarmiento, L. F., Schäfer, S., Schiller, J., Schöfer, P., Schweitzer, A., Solano, E., Stahl, O., Strachan, J. B. P., Suárez, J. C., Tabernero, H. M., Tala, M., Tulloch, S. M., Veredas, G., Vico Linares, J. I., Vilardell, F., Wagner, K., Winkler, J., Wolthoff, V., Xu, W., Yan, F., and Zapatero Osorio, M. R. (2017). The CARMENES search for exoplanets around M dwarfs. First visual-channel radial-velocity measurements and orbital parameter updates of seven M-dwarf planetary systems, *ArXiv e-prints* arXiv:1710.01595 [astro-ph.EP].

Tsiganis, K., Gomes, R., Morbidelli, A., and Levison, H. F. (2005). Origin of the orbital architecture of the giant planets of the Solar System, *Nature* **435**, pp. 459–461, doi: 10.1038/nature03539.

Tu, L., Johnstone, C. P., Güdel, M., and Lammer, H. (2015). The extreme ultraviolet and X-ray Sun in Time: High-energy evolutionary tracks of a solar-like star, *A&A* **577**, L3, doi:10.1051/0004-6361/201526146, arXiv:1504.04546 [astro-ph.SR].

Van Eylen, V., Lund, M. N., Silva Aguirre, V., Arentoft, T., Kjeldsen, H., Albrecht, S., Chaplin, W. J., Isaacson, H., Pedersen, M. G., Jessen-Hansen, J., Tingley, B., Christensen-Dalsgaard, J., Aerts, C., Campante, T. L., and Bryson, S. T. (2014). What asteroseismology can do for exoplanets: Kepler-410A b is a small Neptune around a bright star, in an eccentric orbit consistent with low obliquity, *ApJ* **782**, 14, doi:10.1088/0004-637X/782/1/14, arXiv:1312.4938 [astro-ph.EP].

Vogt, S. S., Marcy, G. W., Butler, R. P., and Apps, K. (2000). Six new planets from the Keck precision velocity survey, *ApJ* **536**, pp. 902–914, doi:10.1086/308981, astro-ph/9911506.

Watson, A. J., Donahue, T. M., and Walker, J. C. G. (1981). The dynamics of a rapidly escaping atmosphere — Applications to the evolution of earth and Venus, *Icarus* **48**, pp. 150–166, doi:10.1016/0019-1035(81)90101-9.

Welsh, W. F., Orosz, J. A., Carter, J. A., Fabrycky, D. C., Ford, E. B., Lissauer, J. J., Prša, A., Quinn, S. N., Ragozzine, D., Short, D. R., Torres, G., Winn, J. N., Doyle, L. R., Barclay, T., Batalha, N., Bloemen, S., Brugamyer, E., Buchhave, L. A., Caldwell, C., Caldwell, D. A., Christiansen, J. L., Ciardi, D. R., Cochran, W. D., Endl, M., Fortney, J. J., Gautier, T. N., III, Gilliland, R. L., Haas, M. R., Hall, J. R., Holman, M. J., Howard, A. W., Howell, S. B., Isaacson, H., Jenkins, J. M., Klaus, T. C., Latham, D. W., Li, J., Marcy, G. W., Mazeh, T., Quintana, E. V., Robertson, P., Shporer, A., Steffen, J. H., Windmiller, G., Koch, D. G., and Borucki, W. J. (2012). Transiting circumbinary planets Kepler-34 b and Kepler-35 b, *Nature* **481**, pp. 475–479, doi:10.1038/nature10768, arXiv:1204.3955 [astro-ph.EP].

Welsh, W. F., Orosz, J. A., Short, D. R., Cochran, W. D., Endl, M., Brugamyer, E., Haghighipour, N., Buchhave, L. A., Doyle, L. R., Fabrycky, D. C., Hinse, T. C., Kane, S. R., Kostov, V., Mazeh, T., Mills, S. M., Müller, T. W. A., Quarles, B., Quinn, S. N., Ragozzine, D., Shporer, A., Steffen, J. H., Tal-Or, L., Torres, G., Windmiller, G., and Borucki, W. J. (2015). Kepler 453 b — The 10th Kepler transiting circumbinary planet, *ApJ* **809**, 26, doi:10.1088/0004-637X/809/1/26, arXiv:1409.1605 [astro-ph.EP].

Whitmire, D. P., Matese, J. J., Criswell, L., and Mikkola, S. (1998). Habitable planet formation in binary star systems, *Icarus* **132**, pp. 196–203, doi:10.1006/icar.1998.5900.

Williams, D. M. and Pollard, D. (2002). Earth-like worlds on eccentric orbits: Excursions beyond the habitable zone, *International Journal of Astrobiology* **1**, pp. 61–69, doi:10.1017/S1473550402001064.

Williams, J. P. and Cieza, L. A. (2011). Protoplanetary disks and their evolution, *Annu. Rev. Astron. Astrophys.* **49**, pp. 67–117, doi:10.1146/annurev-astro-081710-102548, arXiv:1103.0556 [astro-ph.GA].

Wittenmyer, R. A., Horner, J., Marshall, J. P., Butters, O. W., and Tinney, C. G. (2012). Revisiting the proposed planetary system orbiting the eclipsing polar HU Aquarii, *MNRAS* **419**, pp. 3258–3267, doi:10.1111/j.1365-2966.2011.19966.x, arXiv:1110.2542 [astro-ph.EP].

Wöllert, M., Brandner, W., Bergfors, C., and Henning, T. (2015). A Lucky Imaging search for stellar companions to transiting planet host stars, *A&A* **575**, A23, doi: 10.1051/0004-6361/201424091, arXiv:1507.01938 [astro-ph.EP].

Zombeck, M. V. (2006). *Handbook of Space Astronomy and Astrophysics* (Cambridge University Press, Cambridge, UK).

Zucker, S., Mazeh, T., Santos, N. C., Udry, S., and Mayor, M. (2004). Multi-order TODCOR: Application to observations taken with the CORALIE echelle spectrograph. II. A planet in the system HD 41004, *A&A* **426**, pp. 695–698, doi:10.1051/0004-6361: 20040384.

Index